FAITH IN SPACE

By

FRANK M. CHILDERS

[With a foreword by Apollo Astronaut James Irwin]

CONSECRATED PRESS, MERRITT ISLAND, FLORIDA

FAITH IN SPACE

A Statement Of Man's Feelings About God and Man's Place In The Universe

Library of Congress Catalog Card Number 97-94655
ISBN 0-9660100-0-0

First printing December, 1997

Printed in The United States of America

If unavailable in local bookstores, additional copies of this and other publications by Frank M. Childers may be purchased by writing to the publisher, Consecrated Press, P.O. Box 542346, Merritt Island, FL 32953. Please see order blanks in back of book.

ABOUT THE AUTHOR

Frank Childers is a retired aerospace engineer who writes frequently about the space program. Frank's 1980 *Kennedy Space Center Genesis* proved very popular among his friends at the John F. Kennedy Space Center and at the local libraries. In 1974 Frank wrote the *History of Quality Assurance at Kennedy Space Center* (unpublished).

In 1985 Frank was chosen to write the centennial history af his Church in Cocoa, Florida, which proved very popular in his Church and community. Frank's writings included *The Story of Space Flight and Kennedy Space Center* which he was asked to present to the staff of Space Camp Florida in Titusville, in March 1987.

In 1991 Frank completed a biography called *"UNCLE DOCTOR,"* of Dr. William S. Hughlett, a medical missionary to the Congo for 40 years. In 1993 Frank completed" *THE HISTORY OF THE CAPE CANAVERAL LIGHTHOUSE*, published by the Brevard Museum of History and Natural Science in Cocoa, Florida. Frank is now working on another book, *"SPACEFLIGHT'S GODLY HERITAGE."*

Frank was involved in technical writing projects in the areas of quality assurance, industrial safety, and security for over thirty-five years while employed by the U. S. Army, the National Aeronautics and Space Administration (NASA) and Computer Sciences Corporation.

Frank was listed in the tenth and several succeeding editions of the Marquis' *Who's Who In The South And Southwest.* Frank received the B.S. Degree in Radio Engineering from the Indiana Institute of Technology in 1949 and pursued a 40-year career in government and industry. He is a member of the Florida Christian Writers Conference and the Space Coast Writers Guild, Inc., and attends their conferences and workshops.

CONTENTS

DEDICATION

I Dedicate this volume to the memory of the following astronauts and mission specialists who lost their lives while readying themselves to fly or while "reaching for the stars" on an actual flight:

Theodore C. Freeman: Aircraft accident in 1964
Charles A. Bassett, II: Aircraft accident in 1966
Elliot M. See, Jr.: Aircraft accident in 1966
Edward G. Givens, Jr.: Deceased, 1967
Virgil I. Grissom:
Roger B. Chaffee: – Apollo spacecraft fire in 1967
Edward H. White:
Clifton C. Williams, Jr.: Aircraft accident in 1967
Major Robert Lawrence: Aircraft accident in 1967
Gregory B. Jarvis:
Christa McAuliffe:
Ronald E. McNair:
Ellison S. Onizuka: – Shuttle Challenger mission, 1986
Judith A Resnik:
Francis R. Scobee:
Michael J. Smith:
Manley L. "Sonny" Carter, Jr., Deceased, 1991
James B. Irwin, Deceased, August 8, 1991
Donald K. Slayton (Deke), Deceased, June 13, 1993
Karl G. Henize, Deceased, October 5, 1993
Stuart Roosa, Deceased, December 12, 1994
Charles L. Veach, Deceased, October 3, 1995
Robert Overmyer, Aircraft accident, March 22, 1996

AUTHOR'S NOTE

This book is both a collection and a statement. At first, it's a collection of some of the newspaper, magazine, and personal items which I have collected during more than thirty years of work in the space program, and second, it is a statement of man's feelings about God and man's place in the universe. It covers a broad arc from the encouragement I found in the Bible to personal feelings of astronauts as they took their faith with them and traveled in space or walked on the Moon. My greatest encouragement for this book was found in Psalm 8:

1. O Lord our Lord, how excellent is thy name in all the earth! Who hast set thy glory above the heavens.
2. Out of the mouth of babes and sucklings hast thou ordained strength because of thine enemies, that thou mightest still the enemy and the avenger.
3. When I consider thy heavens, the work of thy fingers, the moon and the stars, which thou hast ordained;
4. What is man, that thou art mindful of him? The son of man, that thou visitest him?
5. For thou hast made him a little lower than the angels, and hast crowned him with glory and honour.
6. Thou madest him to have dominion over the works of thy hands; thou hast put all things under his feet.
7. All sheep and oxen, yea, and the beasts of the field;
8. The fowl of the air, and the fish of the sea, and whatsoever passeth through the paths of the seas.
9. O Lord our Lord, how excellent is thy name in all the earth!

I was fortunate to be numbered with the thirty-seven scientists, engineers, mechanics, technicians, soldiers, and administrative personnel who comprised the original "Missile Firing Laboratory" at Cape Canaveral under the direction of Dr. Wernher von Braun and Dr. Kurt Debus. We launched the first U.S. Army ballistic missile from Cape Canaveral in August 1953, and most of us remained with the unit, where we assumed greater leadership roles when transferred to the newly created National Aeronautics and Space Administration (NASA) in 1958. We then became the "core" organization around which the John F. Kennedy Space Center was formed.

After participating in the launching of Explorer I, and the launching of Alan Shepard and Gus Grissom into space aboard our little Redstone rocket, I became aware of the fact that our team was indeed making history. At that point I began to gather data on what our astronomers, scientists, engineers, and astronauts thought individually about invading this domain we call "space."

The general public was also obviously interested in understanding the need...why invade God's domain, why spend our resources on space when so much needed to be done for the underprivileged, the hungry, here on this planet? Would God really bless us in this venture? These were also my own concerns. Should I be involved in such an undertaking? This question surfaced many times in my mind, but each time I was able to assure myself that I definitely felt a calling to be working in the space program.

As this assurance came to full clarity in my mind, I became more earnest about collecting the thoughts and ideas of others like myself who were involved in the space program.

My main motivation for writing this book is to give a measure of "faith in space" to the astronauts who are venturing out and exercising their own faith in such an adventurous way for all of us to see, and to give them a measure of faith in an orderly universe they face at T-0. The consensus of others before them about whether God has blessed, and continues to bless man's venture into this last frontier will hopefully add to their faith in man's place in space. Another motivation was that maybe in some way it might help our people to understand that our space program is not just a venture of the United States alone. It is a venture of the citizens of the whole planet. Therefore mankind must be good stewards of this special responsibility - this special gift from God. All the ventures into space should be taken with the goal of welding all the nations of the world into a closer bond of endeavor and grand accomplishment as nations work peaceably together to open the frontiers of space. Humanity's future depends on exploration of our solar system and the universe beyond.

The current NASA Administrator, Dan Goldin, appointed by President Clinton, made this statement recently on TV:

> "We are explorers, it's written into our genetic code."

President Kennedy set the mood for the United States' goal in space at Rice University on September 12, 1962, when he said:

> "We set sail on this new sea because there is knowledge to be gained, and new rights to be won, and they must be won and used for the progress of all people. For space science, like nuclear science and all technology, has no conscience of its own. Whether it will become a force for good or ill depends on man, and only if the United States occupies a position of preeminence can we help decide whether this new ocean will be a sea of peace or a new, terrifying theater of war."

If having written this book has helped to encourage future astronauts, mission specialists, and those who work in and support the space program, and if it helps to forward this message from a great leader of the recent past, then to God be the glory for His having given me this task.

ACKNOWLEDGEMENTS

For the inspiration and encouragement that led to the completion of this volume, I thank the late Gordon Harris, former Chief of Kennedy Space Center's Public Affairs Office for his ideas on the subject matter, especially concerning Dr. Wernher von Braun's faith; and my good friend Patricia Phillips formerly of KSC Public Affairs Office for her ideas, encouragement, and insistence that I continue the goal of publishing my book.

I offer my gratitude to the author of GOD POWER IN AFRICA - Rev. Dr. Anthony B. Fadely - for his insight into what this volume should convey to all levels of readers, and for his gift of the title, "FAITH IN SPACE." I also appreciate Dr. Fadely's editorial expertise and gifted knowledge of 'faith in God' as given to all of us through God's Holy Word. I indeed thank my wife, Lois, for her time spent in proof-reading, and for her encouragement in finishing this book. I also offer my gratitude to the late Apollo Astronaut, Colonel James B. Irwin for providing the "FOREWORD" for my book.

I owe a great deal of thanks to Mr. Karl Ronstrom for his excellent work with the pictures that appear. I thank the NASA Centers for faithfully providing all of the photographs and cassette recordings that I requested and for the encouragement I sensed in their response letters.

I want to thank former U.S. Congessman Bill Nelson (crewmember, STS-61C) and the following friends from the space program who took time to review and comment on this book: Grady Williams, John Fike, J.D. Collner, Robin Wagoner, Roy Lealman, Ed and Susie Mathews, and Terry Greenfield; members of the clergy who provided insightful comments: Rev. Donald Mackay and Rev. Dr. Ray Harrison, and self-publishing authors; Cathy Katz and Jamie Pritchett.

FOREWORD
by
Apollo Astronaut James B. Irwin

I feel honored and privileged to be asked by Frank Childers to write a foreword to his book, "Faith In Space." I believe that travel in space really exercises ones faith! You must have faith in equipment, faith in people, faith in yourself and faith in God. "Faith in being sure of what we hope for and certain of what we do not see." Hebrews 11:1.

When I came into the space program I was surprised by the great number of Christians working in the program. There were scientists, engineers, technicians and yes, astronauts. They were not ashamed to share their faith. Dr. Wernher von Braun was a great example. He was a brilliant scientist and engineer and leader. When we lost Dr. von Braun, we lost a great leader. He was able to speak to the scientists, the politicians and the man on the street.

I shall never forget a visit to Segovia, Spain. I visited the beautiful castle of Queen Isabella, spoke to the community and had lunch at Candidos. After lunch, Candido showed me the guest book. He was so proud of what Dr. Wernher von Braun had written: "No Segovia, no Isabella - no Isabella, no Columbus - no Columbus, no America - no America, no Moon. With these words, he won the love of Candido and the love of the people of Segovia. Sometimes I add to Dr. von Braun's words to say "No Moon - No Jim Irwin."

Without the journey to the moon, Frank would not have invited me to write this foreword. All of the twelve men who have been to the moon are believers in God and in His son, Jesus Christ. Apollo was the Greek God of Light. It is very interesting that the name would be given to our moon exploration. We believe that we have made many important discoveries and that we have brought back the pure light of God - His son, Jesus Christ, who said, "I am the light of the world."

Some say there is a conflict between science and religion. I disagree. Religion for me is faith in God - a personal relationship with Him through His son, Jesus

Christ. Science is the understanding of God's basic laws that control the universe. If you accept my definition of religion and science there should not be a conflict. Everyone needs faith in God. The scientist particularly needs faith in God because then his discoveries will be enabled by God and the useful implementation will be directed by God. We all need each other and we desperately need God.

On a recent trip to the Soviet Union, we were able to host a banquet for the cosmonauts and their wives. After I gave my testimony, one of the senior cosmonauts asked if he could say a few words. He spoke at great length with tears in his eyes. He said that he had faith in God and that spiritual things were first priority in his life now. Cosmonauts and astronauts were united in space on Apollo-Soyez; we are now united in our efforts to reach even further into space, we are united in our faith because of space.

The loss of the space shuttle Challenger and its crew was a great loss to America and the world. I still ask myself, "Why?" Why was the decision made to launch on such a day? It was a very cold day. There was ice on the space craft, and yet the decision was made to launch. Were we too proud to delay a few more days? Were we too careless in our execution? I am sure the verdict is that we were very careless. I think there is a very painful lesson to be learned from the Challenger tragedy. We have seen what happens when we become careless and/or proud. Horrible things can happen. I just pray that we have learned from Challenger.

Our faith became much, much stronger as a result of our journey into space. I believe God sent me to the moon, so that I might return to the earth to share the son - His son, Jesus Christ. I have been sharing for twenty years that "Jesus walking on the earth is more important than man walking on the moon." I think that event is the most important thing that has ever happened on our blue planet.

We have seen our home as God must see it, and our lives will never be the same. Enjoy Frank's book "Faith In Space" and I pray that your faith will become stronger through these testimonies from our space program.

CHAPTER 1 - SPACE, MAN'S DOMAIN

I had worked with the Dr. von Braun team since early 1952 and had felt the excitement that many American and German scientists and engineers felt as they finally realized their dream of space flight after transferring to the National Aeronautics and Space Administration (NASA) and we began working on the Saturn - Apollo program.

In a timely manner the Apollo 11 mission attained the national goal, set by President John F. Kennedy in 1961, of landing men on the Moon and returning them safely to Earth within the decade of the 1960's. The date was July 16, 1969; the time of liftoff was 9:32 am EDT. My job placed me in the Launch Control Center(LCC) at the Quality Assurance console for ground instrumentation and information systems. Our launch team had worked many hours on this Saturn V (AS-506) since its rollout to pad 39A on May 20th. My job for the launch countdown was to monitor the work of the NASA, Federal Electric and RCA Quality Assurance personnel as they inspected, witnessed and/or, approved the various Quality Inspection Points (QIP's) during the countdown.

Work progressed smoothly throughout the countdown and no major problems were reported as far as the ground instrumentation and information systems were concerned. The Countdown Demonstration Test performed on June 27th, and the Simulated Flight Test, performed on July 2nd, had been good rehearsals for the launch team, with much confidence gained for both equipment and people.

On July 3rd we drained the fuel tanks and let Armstrong, Collins and Aldrin participate in a dry test. This test further assured a good launch.

We launch team members had to be in place several hours before the launch in order to complete all the steps involved in the countdown procedure. This included securing all of the systems and communication lines from the Mobile Service Structure so that it could be moved back to its parking site a mile away. Part of my job was to be sure that certain preparation steps were completed before and after the move started. This included the nitrogen verification of all the various control and instrumentation enclosures which if left without adequate purging, could cause an explosive situation at the launch pad.

The Crawler Transporter operation took place early so that the Mobile Service Structure move could be completed in a timely manner. During the early hours of July 16th, the tanks of the second and third stages were filled with liquid hydrogen. One of my most important jobs was to assure that hydrogen hazard monitoring was accomplished with great reliability. For all the years of Apollo development, I had been assigned as the Reliability Engineer for the Ground Instrumentation and Information Systems. This included the assignment of reliability requirements for contracts as well as the final assessment of reliability status prior to each launch. My 10-member team for this effort was provided by General Electric Company on the NAS-410 contract out of NASA Headquarters. The outstanding job of reliability analysis and reliability profiling for each early Apollo launch which this team did gave management a good feeling about launching.

Brilliant lights illuminated the launch area and Apollo 11 during the night of July 15th. I left my console once to visit the small area between Firing Room 1 and Firing Room 2, to look out of the huge windows toward the launch pad. I saw the beautiful space vehicle with the astronauts on board, poised to make man's first visit to another celestial body. I prayed to God for a safe and successful flight, a flight for which we all had made such extra effort in preparation, and which our country had been anticipating since President Kennedy established the goal in 1961. We were about to reach that goal and I felt a great deal of pride in my heart and mind as I looked toward launch pad 39A.

As I returned to my console, I joined about 450 other people in Firing Room 1, busy at their display and control consoles. My position was at the front area facing out over the main Firing Room. Our Center Director, Dr. Kurt Debus, and his staff sat just behind me at the higher level where they could see the majestic Saturn vehicle and also look out over the entire Firing Room.

July 16th was ushered in by a beautiful dawn. Everything was moving nicely toward the time of ignition and liftoff. No major problems were reported as I can recall, and the excitement began to build as the minutes and seconds ticked off on our console monitors. All of us in the Firing Room were tied to our positions and could not see ignition at T-8.9 seconds except on the individual monitors and on the large-screen overhead projectors. That little wisp of white smoke indicated that the launch vehicle was about to come to life and begin the epic trip to the Moon.

As I looked around the Firing Room at this point I saw and felt the tension and the sometimes prayerful position of individuals who, like me, were committing the astronauts to God. We always felt most helpless at liftoff because we could no longer do anything for the crew.

Exhaustion held us for a few moments, then cheering began to swell as the vehicle climbed into view for us down below the windows. We could see for just a few seconds. We then looked back at our monitors for a better view. This was the culmination of our almost 10-year labor of love as designs were developed, hardware built and tested, and procedures written for getting the Apollo 11 crew to Tranquillity Base.

These are some of my memories of the Apollo 11 mission and reflect thoughts of the analyses completed and flight calculations performed for setting the Lunar Excursion Module down onto the uncertain surface of the Moon's Sea of Tranquillity. They remain in my mind as something almost impossible to have done; especially when I look up at the Moon from my yard near the place where we performed this outstanding feat for mankind.

In the excitement of this activity I began to ask

myself, "What was man trying to do? Did he think he was God? What business did mankind have invading space like this?" This question was also asked by many others upon this, the occasion of man's first visit to another celestial body.

We must look back in time to form the answers. The following narrative history taken from the Civil Air Patrol publication, *The Dawning Space Age,* pages 17-19, by H.E. Mehrens, gives us some of the answers:

> "Space travel, like air travel, challenged the imagination of man since prehistoric times. It is not surprising that fictitious reports of imaginary flights through space were written even during the early centuries. In A.D.160, a Greek named Lukian (Lucian), wrote of a voyage to the moon. He wrote it in Latin and called it *Vera Historia* (A True History). A German astronomer, Johannes Kepler (1571-1630), writing in Latin, also reported an imaginary trip to the moon. Bishop Francis Godwin, in 1638, wrote a story which he called *The Man In The Moone.* Cyrano de Bergerac, of greater fame as a character in an immortal love story, wrote about a journey to the moon in a vehicle powered by huge powder rockets. Research in Chinese history shows that a mandarin, named Wan Hoo, attempted a journey to the moon with 47 large rockets strapped to the framework of his chair. Edgar Allan Poe, writing *The Unparalled Adventures of One Hans Pfall* in 1835, had his hero using a balloon to carry him into "outer space."
>
> Jules Verne was more scientific in his report of a trip to the moon. In his story, *From Earth to the Moon* , written in 1865, he correctly reported the escape velocity to be seven miles per second (the velocity a body must attain in order to escape the major influence of the earth's gravity).
>
> During the years that some writers and thinkers dealt with the fantasy of space travel, others were concerned with the results of scientific investigations

which contemporary scientists use as a basis for solving problems of rocketry upon which space travel depends.

More than seven centuries ago, Roger Bacon (1242) described the composition of the first rocket propellant-gunpowder. The ingredients he prescribed were saltpeter (sodium nitrate) as the oxidizing agent, with charcoal and sulphur. A technical report on rockets was published in France as early as 1561. Similar reports appeared in both Poland and Germany in the 17th century.

In the year 1687, the English scientist, Sir Isaac Newton, defined the laws of motion which rocket propulsion must take into account. In the year 1810, Mr. W. Moore published in London his book *On the Motion of Rockets.* In the year 1881, a Russian named Kibaldchitch worked out in detail a plan for a flying machine powered by a rocket motor. A German inventor, Hermann Ganswindt, in 1890, drew up a similar set of plans and specifications for a space vehicle. His vehicle would depend on dynamite or other high explosive as fuel for the rocket motor.

In the year 1900, a Russian schoolteacher named Konstantin Edwardovitch Tsiolkovsky described a large piloted rocket which he believed could be used for trips into space. The Smithsonian Institution published the first significant book on rocketry and space travel in 1920, entitled *A Method of Reaching Extreme Altitudes,* by Dr. Robert H. Goddard. In the year 1923, a German mathematician, Professor Hermann Oberth, published a similar book, *By Rocket into Interplanetary Space.* A third edition was called *Ways to Space travel.* It was upon the findings of these three scientists that the present-day rocket applications have been built." (1)

As we look at this early history of ideas for space flight through the principles of rocketry, we may wonder why these scientists and inventors pursued such ideas, while the majority of their peers were content to go along as things were. It might be they were anxious to know more about themselves and their relationship to the Creator. The more they studied, the more they wondered. The universe was so vast, it might be that they were trying to find a place in it for themselves. They might have had questions about nature's origin and what was known about it. Questions like: Did God really make the universe? Does He still make things happen? If the universe is so big, how can God know and care about me? These may be our questions, too.

A long time ago, David, the shepherd boy, was full of questions as he gazed into the starry sky. In a Psalm he tells us how he wondered about the moon and the stars - how they got there - how far away they were? He also wondered about his own importance in comparison to the beauty and vastness of the starlit night, expressed In the 8th Psalm as quoted in the Author's Note, p. iii.

The key statement in the Psalm says, "Thou hast given him dominion over the works of Thy hands." David makes the statement also that God ordained the moon and stars for man to have dominion over. It might be that these scientists and inventors of old had a better glimpse into the mind of the Creator than the ordinary person, just as David had.

We have further assurance that God made the universe in Isaiah, 48:13:

"It was my hand that laid the foundations of the earth; the palm of my right hand spread out the heavens above; I spoke and they came into being." (The Living Bible).

Much has been written about the compatibility of science and religion. The more man studies their compatibility, the more he is convinced that these two disciplines are very compatible. True science and true religion do not conflict with each other if we remember that each has its own way of arriving at answers. This does not mean that science and religion operate in entirely separate worlds and that the questions they answer are unrelated. Science and religion are parts of a whole. They are not separate from one another any more than our minds are separate from our bodies. Science and religion work together to help us understand all we can about our world, our lives, and our relation to the world and to God.

Science can help religion with its task of making us aware of God's handiwork. It cannot tell us whether or not there is a God, but it can help us understand His power and His wonder.

When we look into a microscope or when we examine closely the "world under our feet," we are impressed. Here we get some insight into the wonders of God's creation in all its detail. When we study the human body or the starry heavens, we get a new understanding and appreciation of the greatness of the Creator. To see the earth and all of creation through the eyes of the scientists is to understand more fully the majesty, the power, and the greatness of God. This is why some of our greatest astronomers, scientists, engineers, and astronauts were, and are, deeply religious.

The January/February, 1997 issue of "AD ASTRA," the magazine of the National Space Society, quoted Father John Managua, President of DePaul University, who provided at the George Washington University Space Policy meeting on November, 22, 1996 this witty and insightful look into the question of science versus religion: "Religion should provide a stabilizing force to tame the unknown, whereas science is a destabilizing factor and helps us conquer the unknown."

Father Managua is saying that religion can help science with its task. It is religion that gives meaning to the discoveries of science. These discoveries do not make sense and are of no value apart from mankind and his attempt to find his place in God's universe.

Faith is required in both science and religion. Sometimes, people assume that science deals with what can be proved and religion deals with what cannot be proved.

The real difference between them may be control and predictability. Both in science and in religion, we have to believe in and depend upon what cannot be seen and upon what cannot be proven.

For example, we cannot see a radio wave. We see and hear only what it does; therefore, we believe it is real. Likewise, we believe that we do not have to see God with our eyes to believe He is real.

I have been interested in radio since I was a youth. I was a radio operator in the Army Air Corps in World War II, have been a radio amateur since 1949, and I have a B.S. degree in Radio Engineering. I do not have a problem with believing that radio waves exist. I can see and hear what radio waves do; therefore I believe they are real. Many things that we accept in both science and religion are both unseen and unprovable, yet we believe they are real, and we act as if they are real.

Based on the inspired word of the Psalmist, we can take dominion of space as part of the domain of our God, and be assured of His blessing for us on present and future efforts as He has blessed the past efforts of mankind.

Yes, there has been a steady stream of divinely inspired spokesmen who have carried the torch of knowledge of the universe. These included Nicholas Copernicus, (a Polish Priest), Johannes Kepler, (a Protestant) whose religion was a force within, whose thoughts turned mainly to theology, but who found time to pursue wonderful discoveries in astronomy, mathematics and the laws of nature.

Beyond these outstanding giants we find other great thinkers who put God first in their lives for vision and inspiration, and we can be sure that God will continue to bless our efforts as mankind continues advances into His heavenly realm. After all, He made us for His pleasure and gave us dominion over all of His creation. (Rev 4:11 and Psalm 8:6)

CHAPTER 2 - ASTRONOMERS BELIEVE

Most of us do not learn a great deal about the universe from our public school classes or college classes. We do not learn anything about the evidence of divine order (teleology) of the universe until we study information from instrument extensions of man out there sailing along through space. Our early information about the universe came from the fathers of astronomy; Copernicus, Brahe, and Kepler.

After the Dark Ages, astronomy had a rebirth. It can be followed more closely than any other phase of the Renaissance. Astronomy grew from the work of these three inspired pioneers, Nicolaus Copernicus (1473-1543), Tycho Brahe (1546- ?), and Johannes Kepler (1571-1630).

Copernicus, a Polish priest, carried out his research of the heavens in the shelter of the Catholic church. His priestly training surely caused him to believe in God and he wanted to find the true meaning of life and the design of God's whole creation. He was bold in his ideas of a fixed sun with the planets in a circular orbit around it. He also imagined that the Earth spun on its axis. These were bold theories and would have put him in mortal danger if he had not dedicated his work to the Pope in a beautifully written preface to his book, *Revolution of Celestial Bodies.*

Tycho Brahe, son of a Danish nobleman, was born in 1546 and pioneered in organizing observational astronomy. He was well educated and had boundless energy for his favorite work. His measurement of the positions of stars and planets gave the data needed for the next advancement in space science.

Johannes Kepler, a German, came to Brahe's rescue in needed mathematical expertise, an area of Brahe's weakness. Kepler was born a Protestant, and his religion was a strong force in his character. He was sensitive and emotional, a mystic and a sage. He felt that his salvation lay in the study of the Bible, rather than dogma of the Pope. He felt that the heavens were a book written by God, to be studied as reverently as the Bible. He was known to wait for years on an answer to a problem. When the answer came, he gave God the credit. He became one of Brahe's assistants at Prague, Germany, and was given the task of developing the celestial mathematics necessary to prove some of Brahe's theories. Upon Brahe's death, Kepler took his place as Emperor Rudolph II's imperial mathematician and proceeded with observations that led to the first two laws of planetary motion which our space scientists use today in their accurate calculations for space travel and return to earth.

Law I states: The orbit of a planet is an ellipse with the sun at one focus.

Law 2 states: A line drawn from a planet to the sun sweeps out equal areas in equal times.

What marvelous insight into the mind of the Creator! Kepler looked hard at the relationship between the distances of the planets, and found a certain harmony that God had placed in the heavens. This discovery is sometimes called the "harmony law." With the discovery of this harmonic law, Kepler felt that his destiny had been fulfilled.

Our first question to astronomers might be, "Do you place your faith in the universe by observation alone, or do you seek and accept God's word for your position in this marvelous domain of Space which God says He has given to mankind? Or do you deal with imagination exclusively?"

I believe that much of the astronomer's undertaking is involved with his astonishment and his craving for answers to wonders of the universe and all that Space implies. He, like the cosmologist, might even wonder how it all started.

All of us feel that the universe had, in some sense, a beginning -- that it began at a certain moment in time, and under circumstances that we cannot describe, nor can we imagine the forces that brought the world into being at that moment in time. Was it, as the Bible says, that:

"Thou, Lord in the beginning hast laid the foundations of the earth, and the heavens are the work of thine hands?" (Hebrews 1:10)

Present astronomical theory tells us that the universe was created fifteen billion years ago in a fiery explosion, and in the searing heat of that first moment, all the evidence needed for a scientific study of the cause of the great explosion was melted down and destroyed.

A lot has been written about the new story of Genesis that is being called the "Big Bang" theory. This theory has shared media attention with other theories such as the "Solid State" theory which claims that the universe had no beginning and is eternal. New evidence now points to the abandonment of the Solid State Theory by nearly everyone, leaving the Big Bang theory exposed as the only adequate explanation of the beginning.

Other astronomers made important discoveries about the fireball theory of the beginning of the universe. These included Vesto Slipher, who at Lowell Observatory, Flagstaff, Arizona in 1913, discovered that several neighboring galaxies were moving away from the earth at high rates of speed. This gives us the first hint that the universe is expanding.

Following along, William de Sitter, a Dutch astronomer, found a solution to Einstein's equations of general relativity, which were published in 1917, and on this solution he based his belief in an exploding universe, in which the galaxies moved rapidly away from each other. De Sitter, in the next eight years had clocked about 40 galaxies and found that they were moving away from the earth at similar rates of speed.

One of my resources stated that Albert Einstein met Dr. De Sitter during his visit to California in 1932, and entered into lengthy blackboard formulae and discussions with him. Einstein had resisted de Sitter's theory since they first corresponded in 1917, but later his studies of Dr. Edwin P. Hubble's observations on speeds and distances of the galaxies finally convinced him that the theory was correct. Shortly before his death, Einstein told a visitor that he fully accepted the idea of "a beginning."

Einstein was awarded the Nobel prize for his contribution to quantum theory. He never accepted the idea that the universe was governed by chance which quantum theory proposes. His feelings were summed up in his famous statement; "God does not play dice with the Universe."

As Dr. de Sitter left his studies of the galaxies to enter other fields, two other astronomers took up his studies. They were the Americans, Edwin P. Hubble and Milton Humason. Hubble concentrated mostly on speeds of galaxies and Humason measured their distances. Using Humason's measurements of distance, Hubble plotted his derived speed against them and arrived at an amazing relationship known now as Hubble's Law: The farther away a galaxy is, the faster it moves. This gave mankind further proof of an expanding universe which underscores Einstein's belief in "a beginning." Hubble's Law is the foundation of the scientific corroboration of Genesis.

My good friend, Gordon Harris, the man known as the Voice of NASA during the 1960's and 70's, wrote a piece for the *TODAY Newspaper* concerning NASA research that touched on the subject of creation, but I take the following from the same reference used; the February, 1978 issue of NASA Activities:

> "The Big Bang theory which attempts to explain creation of the universe faces a new challenge from researchers employing a high-flying NASA aircraft, one of the U-2 spy planes. As a result of their findings, Drs. Richard Muller and George Smoot of the Lawrence Berkeley Laboratories in California believe the cosmos began with a powerful but controlled and completely uniform expansion not unlike the blossoming of a

flower.

Their explanation more nearly fits the concept of fundamental Christianity and the Biblical description of Genesis.

Researching the cosmos microwave background by means of ultra sensitive radio equipment aboard the U-2, the scientists concluded that planet Earth was formed serenely along with other cosmic bodies while matter and energy were uniformly distributed, expanding at equal rates in all directions.

By measuring the movements of the Earth against the 'glow' left from the fireball in which the universe was born, they also found the Milky Way Galaxy, together with the solar system and Earth, appear to be hurtling through space at more than one million miles per hour towards the constellation Hydra.

Marc Gorenstien, a graduate student who took part in the experiments, said radiation from the universe-forming event 15 billion years ago is so uniform that it provides a universal reference for measuring this motion.

Smoot added that the data did not show any rotation of the universe. 'If there is rotation,' he added, 'it has to be less than one hundred millionth of a rotation in a billion years. Yet everything within the universe is rotating planets, stars, and galaxies."

According to their report, 'the big bang appears finely orchestrated. Either conditions before the beginning were very regular, or processes we don't yet know worked to make the universe extremely uniform.'

'Big Bang' proponents believe the universe began as a hot, incredibly dense mass containing all the matter in the cosmos. The primeval fireball exploded in an immense cataclysm.

According to this, protons and neutrons

began to fuse into nuclei; at first hydrogen, then deuterium and later helium evolved. After millions of years, the material cooled to condense into galaxies, stars and planets.

'The large-scale regularity we found in the expansion of the universe makes the million-mile-an-hour random local motion of the Earth and our galaxy all the more surprising,' Smoot said.

Dr. Muller added that the U-2 instruments measure radiation just reaching Earth from the remnants of creation farthest removed from our planet. 'Their light is only now reaching us,' Muller explained, just as our light is now reaching them." (2)

Another proponent of an orderly universe is Professor Stephen W. Hawking who holds Newton's Chair as Lucasian Professor of Mathematics at Cambridge University. Professor Hawking is widely regarded as the most brilliant theoretical physicist since Einstein. In his book, *A Brief History of Time*, Bantum Books, April 1988, he discussed some possible creation singularities and gave attention to the laws of gravity which shape the large-scale structure of the universe. He made the statement that at the Big Bang and other singularities, all universal laws would have broken down, so God would still have had complete freedom to choose what happened and how the universe began.

Upon reviewing the March/April 1990 issue of NASA Activities, I was greatly pleased to see that NASA's Cosmic Background Explorer (COBE), launched in the fall of 1989, had documented and presented some exciting news about early creation of the universe:

"Preliminary results are in accord with the predictions of the Big Bang theory, which traces the origin of the universe to a primordial explosion some 15 billion years ago. The universe today shows that sometime after the Big Bang additional release(s) of energy occurred. COBE's new results severely limit the magnitude and character of such a release. Limited COBE data now indicate a smooth, uniform Big Bang." (3)

This is no surprise to many because the believer can see God's hand more clearly in creation after this data was recorded by COBE. We should look forward to more startling facts about early creation as space science progresses and knowledge is accumulated.

Each time I see words used in these reports such as: Serene, Uniformly Distributed, Extremely Smooth, Finely Orchestrated, Extremely Uniform, Very Regular, I can think only of God. He is the only one who could have caused such beautiful orderliness and control of creation, and science continues to surprise us and make us glad to be worshiping Him.

The Goddard Space Flight Center was very cooperative in providing recent data from their COBE mission in the form of color graphs and scientific data. Much of their data was explained in their data request coordination letter of October 6, 1993:

...........Data from the COBE experiment were never intended to address the creation of the universe from a religious standpoint. Scientific experiments, including those conducted by the COBE instruments, tell us about how the universe developed from the time it was a tiny fraction of a second old to its present state. COBE observations give further definitive credence to the Big Bang theory, and for the first time, COBE detected minuscule ripples in the "temperature" of the cosmic microwave background radiation. These ripples are believed to be the primordial tenuous concentrations of matter that condensed and, over the course of some 15 billion years, developed into the various structures seen throughout the universe today: stars, galaxies, clusters of galaxies, and the vast empty spaces, called "voids," between those clusters. Previous experiments were not sufficiently sensitive to detect the ripples.

The following chart provided by the Goddard Space Center's Office of Public Affairs shows six all-sky maps that were constructed from preliminary data from the Differential Microwave Radiometer instrument aboard the COBE spacecraft. The Goddard scientists explained the maps as follows:

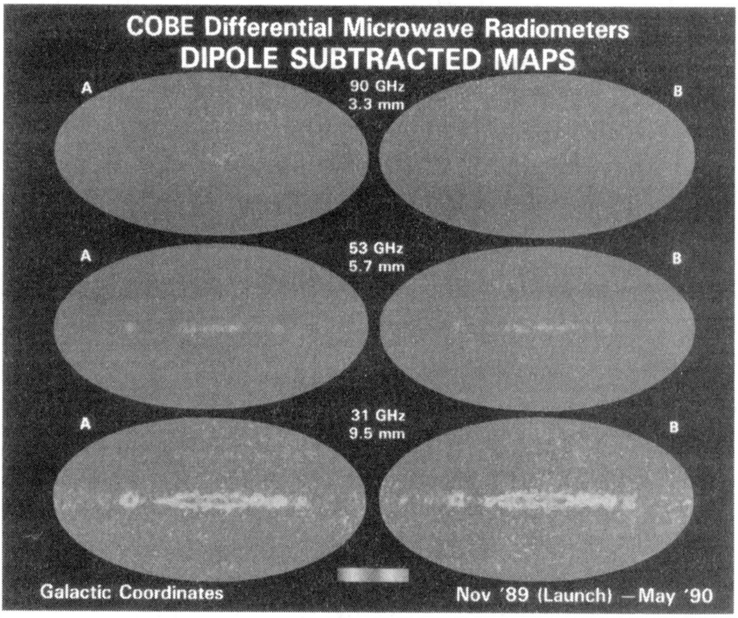

COBE Microwave Sky Maps (Provided by Goddard Space Flight Center)

"The maps show the appearance of the sky at three microwave frequencies after the motion of the Solar System has been removed. There are two independent channels, marked "A" and "B" at each of the three frequencies--31.5, 53, and 90 GHz. The maps are shown in galactic coordinates, with the plane of the Milky Way Galaxy horizontal across the middle and the galactic center at the center. The motion of the Solar System relative to the distant matter is known as the dipole anistrophy and has been subtracted from the maps. The Milky Way Galaxy can be seen in all the maps. It is strongest at 31.5 GHz and weaker, but still present, at 53 and 90 GHz. These maps show that the early universe was very smooth. No "lumpiness" in the early universe has been detected down to one part in 25,000 with these latest results. Fluctuations in the maps are consistent with instrument noise in this preliminary data set. Data from six months of observations were used to make these maps. COBE, designed to study of remnants of the "Big Bang," was launched in November 1989."

A special phenomenon, I am learning, is the extreme orderliness in the universe that we perceive or understand from the studies of God's spokesmen at their telescopes, scientific spacecraft, and at their blackboards.

Through the Associated Press, on June 11,1993, the *Sky & Telescope* magazine, announced a contest to rename the Genesis of the universe, stating that: "The 'big bang' is a catchy phrase that conjures up the wrong type of creation." University of California astronomy teacher Timothy Ferris was appointed judge for the contest. Timothy said this about the whole issue, "The big bang theory says all time and space was contained in the expansion. There doesn't have to have been anything violent about it."

Even the eminent agnostic, Thomas H. Huxley, confesses himself: "utterly unable to conceive matter if there were no Mind to feature that existence." In his book, *Science And Hebrew Tradition*, page 47, Huxley made this statement:

> "But once originated, the conception of the constancy of the order of Nature has become the dominant idea of modern thought. To any person who is familiar with the facts upon which that conception is based, and is competent to estimate their significance, it has ceased to be conceivable that chance should have any but the natural sequence of cause and effect. We have come to look upon the present as the child of the past and as the parent of the future; and, as we have excluded chances from a place in the universe, so we ignore, even as a possibility, the notion of any interference with the order of Nature. Whatever may be man's speculative doctrines, it is quite certain that every intelligent person guides his life and risks his fortune upon the belief that the order of Nature is constant, and that the chain of natural causation is never broken." (4)

Just how orderly is nature and can one depend upon it? As an illustration let me share an amazing story that was circulated by Christian prayer groups in the National Aeronautics and Space Administration as having happened at Greenbelt, Maryland:

"Did you know that the space program is busy proving what has been called <u>myth</u> in the Bible to be true? Mr. Harold Hill, president of the Curtis Engine Company in Baltimore, Maryland, and a consultant in the space program, relates the following development:

> "I think one of the most amazing things that God has for us today happened recently to our astronauts and space scientists at Greenbelt, Maryland. They were checking the position of the sun, moon, and planets in space in regard to where they would be 100 years and 1,000 years from now.
>
> We have to know this so that we don't send a satellite up and have it bump into something late on its

orbits. We have to lay out the orbit in terms of life of the satellite, and where the planets will be so the whole thing will not bog down. They ran the computer measurements back and forth over the centuries and it came to a halt.The computer stopped and put up a red signal, which meant that there was something wrong either with the information fed into it or with the results as compared to the standards. They called the service department to check it out and they said, 'It's perfect'. The IBM head of operations said, 'What's wrong'? 'Well, we have found there is a day missing space in elapsed time.' They scratched their heads and tore their hair. There was no answer!

One religious fellow on the team said, 'You know, one time I was in Sunday School and they talked about the sun standing still.' They did not believe this, but they didn't have any other answer so they said, 'Show us.' So he got a Bible and went back to the book of Joshua where they found a pretty ridiculous statement for anybody who had 'common sense'. There they found the Lord saying to Joshua; 'Fear them not: for I have delivered them into thy hand; there shall not a man of them stand before thee.' Joshua was concerned because he was surrounded by the enemy and if darkness fell, they would overpower them. So Joshua asked the Lord to make the sun stand still. That's right! 'The sun stood still and the moon stayed, and hasted not to go down about a whole day.' The space men said, 'There is the missing day.' Well, they checked the computers going back into the time it was written and found it was close, but not close enough! The elapsed time that was missing back in Joshua's day was 23 hours and 20 minutes, not a whole day.

These little words in the Bible are important but they were still in trouble because if you cannot account for 40 minutes, you'll be in trouble 1,000 years from now. Forty minutes had to be found because it can be multiplied many times over its orbits. Well, this religious fellow remembered also somewhere in the

Bible where it said the sun went backwards. The space men told him he was out of his mind. But they got out the Bible and read these words in II Kings, chapter 20: "Hezekiah, on his death bed, was visited by the prophet Isaiah, who told him that he was not going to die. Hezekiah did not believe him and asked for a sign as proof. Isaiah said, 'Do you want the sun to go ahead ten degrees?' Isaiah spoke to the Lord and the Lord brought the shadow Ten Degrees backward. Ten degrees is exactly 40 minutes! Twenty-three hours and twenty minutes in Joshua, plus 40 minutes in II Kings, make the missing 24 hours the space men had to log in the log-book as being the missing day in the universe. Isn't that amazing? Our God is rubbing their noses in His truth! That's right!"

My copy of this amazing story has this notation at the bottom: (From the University Park Church of Christ, Hyattsville, Maryland).

Another account of this story begins on page 280 of Dr. Harry Rimmer's book, *The Harmony of Science and Scripture* . This account is included with permission of W.B. Eerdmans Publishing Company:

"Sir Edwin Ball, the great British astronomer, found that twenty-four hours had been lost out of solar time. Where did that go, what was the cause of this strange lapse, and how did it happen? The answer may be expected in vain from sources of human wisdom and learning! There is a place, however, where the answer is found. And this place is attested by a scientist of standing. There is a book by Prof. C.A. Totten of Yale, written in 1890, which establishes the case beyond the shadow of a doubt. The condensed account in his book, briefly summarized, is as follows:

'Professor Totten wrote of a fellow-professor, an accomplished astronomer, who made the strange discovery that the earth was twenty four hours out of schedule! That is to say, there had been twenty-four hours lost out of time. In discussing this point with his fellow-professors, Professor Totten challenged this man to investigate the question of the inspiration of the Bible. 'You do not believe the Bible to be the Word of God, and I do. Now here is a fine opportunity to prove whether or not the Bible is inspired. You begin to read at the very beginning and read as far as need be, and see if the Bible can account for your missing time.'

The astronomer accepted the challenge and began to read. Some time later, when the two men chanced to meet on the campus, Professor Totten asked his friend if he had proved the question to his satisfaction. His colleague replied, 'I believe I have definitely proved that the Bible is not the Work of God. In the tenth chapter of Joshua, I found the missing twenty-four hours accounted for. Then I went back and checked my figures, and found that at the time of Joshua there were only twenty-three hours and twenty minutes lost. If the Bible made a mistake of forty minutes, it is not the Book of God!'

Professor Totten said, 'You are right, in part at least. But does the Bible say that a whole day was lost at the time of Joshua?' So they looked and saw that the text said, 'about the space of a whole day.' The word 'about' changed the whole situation, and the astronomer took up his reading again. He read on until he came to the thirty-eighth chapter of the prophet Isaiah. In this chapter, Isaiah has left us the thrilling story of the King Hezekaiah, who was sick unto death. In response to his prayer, God promised to add fifteen more years to his life. To confirm the truth of His promise, God offered a sign. He said, 'Go out in the court and look at the sundial of Ahaz. I will make the shadow on the sundial back ten degrees!' Isaiah recounts that the king looked, and while he looked, the shadow turned backward ten degrees, by which ten degrees it had already gone down! This settles the case, for ten degrees on the sundial is forty minutes on the face of the clock!

So the accuracy of the Book was established to the satisfaction of this exacting critic.

When the astronomer found his day of missing time thus accounted for, he laid down the Book and worshipped its Writer, saying; Lord, I believe!" (5)

The reason I wanted to share these stories with the reader is to emphasize the orderliness of the Creator's Universe. If this missing time could not be accounted for, the celestial mathematics and mechanics which allows the engineers and computers to place satellites into specified orbits, and to send spacecraft into distant space would not be accurate enough to depend on. Also we could not have orbited the moon, much less have placed men and spacecraft on the moon in predicted places.

There is even some orderliness about the known comets, of which Halley's comet is the favorite. Sir Edmund Halley, for which this comet is named, believed there was orderliness in the Universe and that there was a God in spite of some of his actions and earlier writings. Research into his life revealed that his interests knew no limits. He made important contributions in the fields of astronomy, physics, mathematics, geophysics, archeology and navigation. He found a way for sailors, especially in the field of navigation, to determine longitude at sea. Halley developed one of the first diving bells and was the first man to chart the wind patterns of the world. His combined knowledge of the sea and of the heavens resulted in his successful linking of the ocean tides to the positions of the moon.

After much acclaim as a navigator, Halley was appointed a captain in the British navy. On one of his voyages, he survived mutiny, typhoid, and icebergs of the Antarctic. The icebergs almost caused his ship to wreck. Upon successful navigation, calm seas, and a proper wind he praised God for saving his ship and crew.

The December, 1978 issue of *Christian Herald* included an interview with Dr. Owen J. Gingerich, the distinguished professor of astronomy and of the history of science at Harvard University. Dr. Gingerich also served as the astrophysicist of the Smithsonian Astrophysical Observatory at Cambridge, Massachusetts. As a scientist and science historian, Dr. Gingerich's research interests ranged widely, from the recomputation of an ancient Babylonian mathematical table to the interpretation of the solar ultraviolet spectrum. He was an authority on subjects as diverse as the atmosphere of the sun, the 17th century astronomer Johannes Kepler and the 16th century astronomer Nicolaus Copernicus.

Dr. Gingerich was the author of books and numerous articles, and he served on the advisory committee of the Journal of the American Scientific Affiliation, a professional association of Christians working in many scientific disciplines. He was active in the Mennonite fellowship. A portion of the interview which appeared in the December, 1978 issue of the Christian Herald Magazine, follows by their permission:

You came to your scientific career as a person of Christian faith, is that correct?

"That's correct. I had a long upbringing within the Mennonite tradition, a tradition that in our family goes back many generations. My father was an eminent figure, a historian in the Mennonite Church. From him I learned things both about faith and about scholarship in a Christian context."

You said that you believe with the Psalmist that the heavens declare the glory of God. How are you finding the heavens declaring His glory?

"I always keep looking for bits of teleology, evidences of divine order, that are found in science. One of the most remarkable, for example, is the internal structure of the carbon, beryllium and oxygen atoms - the detailed structure of what are called the resonance levels of the atomic nucleus. If the resonance level in carbon were placed one-half a percent lower, the nuclear reaction that creates carbon from beryllium would be virtually impossible. There would be practically no carbon in the universe and therefore no possibility of life, because the carbon atom is indispensable for building complex organic molecules from which life is made. Similarly it goes just the other way in the oxygen atom, There, if the resonance level were half a percent higher, the carbon would have been driven away into oxygen in the process of nuclear building in supernovae, and again carbon would be very rare. The famous English cosmologist, Fred Hoyle, is reputed to have said that nothing shook his atheism quite as much as his noticing this detail of the atomic structure. One has two possible ways of looking at it: Either it's an incredible coincidence, or if you look at the universe with a different set of spectacles, it's an obvious way that the heavens declare the Glory of God. That's why I say that you can't simply prove these things and have people come away from an astronomy course convinced in a belief in God. But on the other hand if you have some other means of taking the leap in faith, then things hold together very well.

Recently I was in Boulder, Colorado, for a fascinating conference on the variability of the sun in past ages - historical ages, prehistorical, and even on a longer time scale. One of the remarkable elements of this discussion was the following interesting situation:

'We think that our sun has been shining for about five billion years. But it hasn't been shining steadily during that time with the same brightness. That is to say, our present understanding of the way in which stars change suggests that originally our sun was only 70 percent as luminous as it is at the present time. In the past few years astronomers have been able to model the behavior of our atmosphere, and it's very interesting to notice that if the mean temperature of the surface of the earth would drop only a small fraction of a degree, the earth would be a great deal colder. If you drop the temperature just a few degrees more, we would get another ice age. At the present time, if the luminosity of the sun would drop by 20 percent, this would plunge the earth into a permanent glaciation. The oceans would freeze solid to the very bottom, and life would be absolutely impossible on our planet.

Now, if our ideas of the sun and its past behavior are correct, then we have a problem on our hands. How could the sun have been so much cooler, and yet not have the earth show any evidence of being frozen solid? It isn't frozen solid now, but it should be, if the sun was that faint and the earth always had its present atmosphere. But it turns out that the atmosphere was not like this in the beginning. It was made primarily of hydrogen. But there was also a great deal of carbon dioxide, and a certain amount of water vapor. And these gases produced what is called a greenhouse effect. The sunlight could penetrate through them, but as the earth and its surface were heated and they reradiated this heat in the infrared, these gases were opaque to the infrared and the light couldn't get out. That's the same thing that happens in a greenhouse, which is why a greenhouse can be warm inside, even on a cold day. The glass is opaque to the infrared that's trying to radiate back out again. Apparently the greenhouse effect kept the earth from freezing solid in very far bygone times. However, this presents a certain problem also. As the sun brightened, the earth should have gotten hotter and hotter.

We can see this in the planet Venus, which is hundreds of degrees at the surface, because it has a carbon dioxide atmosphere and a super greenhouse effect. There isn't any living material on the planet Venus. So why was it that our earth changed around, and switched its atmosphere? As far as we can tell, it was the emergence of life that began to use up the carbon dioxide and fix it, and hydrogen-using kinds of animals at the beginning which then slowly changed our atmosphere over to the present nitrogen-oxygen mixture. And it's quite remarkable that this switchover occurred at just the right moment in the history of the earth, when everything was precariously balanced. Had it happened too soon, the oceans would have still been frozen and this would now be a dead, icy planet.

On the other hand, if the change hadn't happened quickly enough, a runaway greenhouse effect would have taken over, and we would have again a lifeless planet with a temperature of hundreds of degrees at its surface. So I tell my class at Harvard, not only do we have here in the earth a unique kind of environment, which is in a vast emptiness of space, but a globe unique in the historical circumstances that have taken place, so that we find a particularly rich surroundings that has survived in spite of a great deal of precariousness.'

How do scientists feel when they come up with findings like those?

'I thought it was interesting at the conference in Boulder that the person who described the remarkable suddenness with which the atmosphere changed at just the right moment was extremely ill at ease about it, because it seemed to him too miraculous. He was uneasy about this as something in science that shouldn't be this way. And of course from my perspective that's all very fine. I would say; 'Well, that's part of a divine plan of the way things worked out. We can't always see things this clearly." (6)

This gives us some insight into the lives and beliefs of the astronomers of the ages who really did establish the basic mathematics, astronomy, aeronautics, astronautics, and geophysics involved in understanding God's Universe and understanding why mankind should assume dominion over "His Works." It is conceivable that God raised up these great scientists and thinkers to be His spokesmen in their time just as He had chosen Prophets like Isaiah and Jeremiah, and His servants such as David and Moses. Does God raise up spokesmen today to carry out His plans for His creation? We can be sure that He does, and many of them are renowned astronomers, scientists, engineers and astronauts, who are living and working in their chosen fields with a deep faith in God and a great faith in God's orderly universe.

CHAPTER 3 - SCIENTISTS BELIEVE

In science, as in religion, there are some things we have to accept without proof. These are called "basic assumptions." They cannot be proven, yet we have to believe in them and act as if they are true. The May, 1968 issue of the *Reader's Digest* had an article condensed from *Redbook* , by Dr. Warren Weaver, a distinguished mathematician and author of numerous interpretive articles and books about science who had this to say as he answered the question, "Can a Scientist believe in God?:"

> "I have given a great deal of thought to this question, and am completely convinced that religion and science are not enemies. I believe that the happy compatibility of science and religion is achieved in much the same way that a marriage is made successful; by each side devoting time and care to learning enough about the other to understand the other's point of view, and by each side respecting the other. We should start with an understanding of what science is. Put most simply, it is the activity by which man gains in his understanding and control of nature. Science is very good at answering the question, how? Religion is good at answering the question, why?" (7)

The October, 1960 issue of the *Reader's Digest* published an article entitled, "Seven Reasons Why a Scientist Believes in God", by A. Cressy Morrison, former president of the New York Academy of Sciences, that we should look at in order to understand why learned men and women choose to believe in, and have a deep faith in the Creator. Mr. Morrison counts:

> First: By unwavering mathematical law we can prove that our universe was designed and executed by a great engineering intelligence.
>
> Second: The resourcefulness of life to accomplish its purpose is a manifestation of an all-pervading

Intellegence.

Third: Animal wisdom speaks irresistibly of a good Creator who infused instinct into otherwise helpless little creatures.

Fourth: Man has something more than animal instinct - the power to reason.

Fifth: Provision for all living is revealed in such phenomena as the wonders of the genes.

Sixth: By the economy of nature, we are forced to realize that only infinite wisdom could have foreseen and prepared with such astute husbandry.

Seventh: The fact that man can conceive the idea of God is in itself an unique proof.

Mr. Morrison continues:

> The conception of God rises from a divine faculty of man, unshared with the rest of the world; the faculty we call imagination. By its power, man and man alone can find the evidence of things unseen. The vista that power opens up is unbounded: indeed, as man's perfected imagination becomes a spiritual reality, he may discern in all the evidences of design and purpose the great truth that heaven is wherever and whatever; that God is everywhere and in everything but nowhere so close as in our hearts. It is scientifically as well as imaginatively true, as the Psalmist said: "The heavens declare the glory of God and the firmament showeth His handiwork!" (8)

When we think of great scientists who have led the world in the conquest of space we think first of our own Dr. Robert H. Goddard, who, by 1926, had constructed and tested successfully the first liquid fuel rocket. He was indeed a world rocket pioneer and the father of modern rocket propulsion in America. He and Konstantin Edwardovich Tsiolkovsky, of Russia and the great German professor, Hermann Oberth, can accurately be called our "fathers of space travel."

We will now look into the lives and accomplishments of these three fathers of space travel and into the life of one space scientist in particular who devoted his whole life to implementing their dreams and designs for space travel.

Dr. Robert H. Goddard

Dr. Eugene M. Emme, NASA Historian, wrote a short sketch on the historic role that Dr. Robert H. Goddard played in the science and technology of rocket propulsion in America. I quote from that sketch:

> "A physicist of great insight, Goddard also had an unique genius for invention. Not only realizing the potentialities of space flight, Goddard constructed, static tested, and flew rockets to confirm his propulsion theories and engineering solutions.
>
> This rare talent in both creative science and inventive engineering places Goddard high among the rocket pioneers. Yet his scientific papers were copiously studied, his speculation and experiments ridiculed, and his genius largely unappreciated in his own day. Little wonder that this modest man once called "the Moon-Rocket Man" and "Moony" has only recently been fully recognized. By 1926, Goddard had constructed and tested successfully the first rocket using liquid fuel. Indeed, the flight of Goddard's rocket on March 16, 1926, at Auburn, Massachusetts, was a feat as epochal in history as that of the Wright Brothers at Kitty Hawk. Yet, it was but one of Goddard's "firsts" in the now booming significance of rocket propulsion in military missilery and the scientific exploration of space.
>
> Primitive in its day as the achievement of the Wrights, Goddard's rocket success made little impression upon responsible government officials. Modest grants from the Smithsonian Institution, and the larger grants from the Daniel and Florence Guggenheim Foundation, as well as protracted leave of absence from Clark University (where he was head of the Physics Department) largely enabled Goddard to sustain his lifetime of devoted research, development and flight testing.
>
> Eighteen years after his successful demonstration at

Robert H. Goddard, with the first successful liquid-fuel chemical rocket, launched March 16, 1926 (NASA Photo)

Auburn, Goddard's basic concepts and many of his technical designs came to world-wide notice in the German V-2 ballistic missile. The advent of the intercontinental missiles, earth satellites and spacecraft was not only based upon that which Robert H. Goddard devoted his creative talents, but also opened up a new era in the accelerating impact of science and technology upon the affairs of mankind."

In closing this historical sketch of our own "Father of Space Travel", Dr. Emme continued :

"In hindsight, Dr. Robert H. Goddard was one of the first scientists who realized the potentialities of space flight and missiles. But he also brought them to a practical state of development.This rare creative and practical talent was, as Mrs. Goddard points out, combined with his dogged persistence and methodical scientific dedication resulted in more than 200 patents on rockets and rocket apparatus issued to Robert H. Goddard, many after his death.

The labor of this scholarly modest man was appreciated by experts but went largely unrecognized until the recent dawn of what is now called the "space age." High honors and wide acclaim, belated but richly deserved, now come to the name of Robert H. Goddard. On September 16, 1959, the 86th Congress authorized the issuance of a gold medal in his honor. Goddard dinners and awards are annual events of the American Rocket Society, the National Rocket Club, and the Air Force Academy. On June 28, 1960, the Smithsonian bestowed on Goddard its highest award, The Langley Medal. And, on July 13, 1960, the American Rocket Society at last erected a fitting memorial on the site of the world's first liquid-fuel rocket flight of 1926 at Auburn, Massachusetts.

NASA's new Goddard Space Flight Center, with its important responsibilities, is a living and working tribute to his historic contributions in the exploration and investigation of space. Once his biography, notebooks, and papers are fully available, Robert H. Goddard's trail-blazing role in the theory and application of rocket propulsion will be even more greatly appreciated."(9)

From theory through laboratory demonstration, through design, construction, test flight, and use of payload, Robert H. Goddard must rank as the U.S. pioneer in modern rocketry. The famous photograph on page 31 shows Goddard beside his first successful liquid-fuel rocket, flown March 16, 1926. Through thick and thin, through good times and bad times like the depression in the United States, and through times of financial support from the Smithsonian Institute and the Guggenheim Foundation and times of no support, this great but humble scientist continued his research into rocketry and its use in putting man into space. What was his driving force? In the book by Milton Lehman entitled, *This High Man*, we find a clue from a statement by Henry Breckinridge, the Guggenheim Foundation attorney when he said: "You couldn't possibly stop him. He was indomitable. He had restraint, self-respect and a faith that would move mountains. He behaved as if he had all the time he needed to carry out his work."

Dr. Goddard came from an Episcopalian background. His father was a devout Episcopalian and Dr. Goddard and his wife, Esther, were married in the Episcopal Church. While they were at their research station near Roswell, New Mexico, a biographer reported that Dr. Goddard was asked to address the youth group of St. Andrews Episcopal Church which they occasionally attended. His topic was on the relationship between science and religion. He told the youth that there was no conflict between science and religion. He told them that science was knowledge and that religion was right living and acting. He said that to keep the world going, we need both of these forces, working hand in hand. He advised them to not be passive because evil and crime are active and aggressive, and he advocated militant decency throughout the world.

Dr.Goddard's financial support from the Guggenheim Foundation would not have been possible without the enduring support and influence of Charles A. Lindbergh. As Goddard would complete one phase of his research and begin another, he could depend on Lindbergh to arrange more financing for his efforts in rocket research. Lindbergh showed a great interest in the research and tried to get Dr. Goddard to go more public with his findings, but to no avail. Dr. and Mrs. Goddard were really impressed with the young Lt. Lindbergh who had flown the Atlantic Ocean, solo, to the inspiration of the whole world.

I know how they felt, because Lindbergh visited one of our rocket launch pads at Cape Canaveral during the time we were preparing to launch one of our Army Redstone missiles. I was busy with my duties at Complex 5/6 as the engineer in charge of vehicle measurements, and for the blockhouse instrumentation equipment. My duties required me to be there the day Charles Lindbergh came to visit, and I shall never forget the experience of meeting him and observing his interest in what we were doing. He was a tall, fine looking man with a sense of respect for the hardware we were using to check out the rocket. He lingered around the launch pad and blockhouse for a while, and it gave us a real lift that such a celebrity could find time to visit us. At the time we did not know of his support of Dr. Goddard's research.

This nation owes a debt of gratitude to Charles A. Lindbergh for his constant support of Dr. Goddard's research. And of course we owe a debt of gratitude to their respective wives. Mrs. Esther Goddard gave constant support to her husband during all the years of their marriage as she assisted him in his research, in his record keeping, and in acting as his official photographer during the static testing and launching of his rockets. Even after his death, she continued the work of assembling his records and notes for the world to have, and pursued several of his patents until they were granted. As for Mrs. Anne Morrow Lindbergh, we can be assured by history that she gave unrelenting support to her hero husband in all his endeavors concerning the science of flight by both airplanes and rockets. I want to ackowledge Mrs. Lindbergh, for in her book *EARTH SHINE*, Harcourt, Brace, & World, Inc. New York, pages 33-37, she chronicled the astronauts of Apollo 8 as they read the creation story during their mission around the moon and back to earth. On page 40 she mentioned that; "What lifts our hearts today seems to be more in the realm of the human, the psychological, and the spiritual. Perhaps, as Konrad Lorenz suggests, 'space exploration safely absorbs man's aggressive and competitive instincts, and in applauding the astronauts' exploits, we are grasping at a hope of preserving peace on earth.' Those noble qualities of man -- heroism, self-sacrifice, dedication, comradship in a common cause - which are tragically brought out in war, are evoked in many phases of the space development. And these qualities must continue to be aroused in some fashion, or man will cease to be all that man can be."

To the consternation of the Guggenheims and the Smithsonian folks, Dr. Goddard remained by himself in his inventive tasks of rocket research. As Europe became more embroiled in the advancing World War II, he began to think of ways his research could aid the United States in case it became involved. He tried contacts within the Army Air Corps and the Navy. He even made trips to Washington to meet with authorities. In each instance he met with little success.

These fruitless attempts to interest the U.S. authorities in his work did not diminish his interest in future advances in rocketry and in the rocket's use as a propulsion vehicle for space travel.

He envisioned earth-orbiting satellites and interplanetary and even interstellar flight. Research into his life's story show that he even anticipated a form of atomic energy for space flight even before the atomic research was begun. He envisioned solar power as a means of propulsion, using mirror-like sails or panels like our satellites use today.

He was concerned that one day, far in the future, the sun and its planets might grow cold, or lack life-sustaining resources for man's survival. At that time, men might become universal explorers and seek more hospitable grounds where they might live and propagate the human race.

These marvelous ideas seem inspired of God; as if He had planted such ideas and dreams of man's ultimate accomplishments in this humble professor. Today, in the 90's, we can claim some of his vision as reality in our day. Much has been done to advance man's conquest of space including the landing on the moon on July 20, 1969. World War II was definitely a driver in building upon Dr. Goddard's research. The fields of medicine and aerospace technology have brought to fruition many of his dreams and predictions in just a few years. Who knows what God has in store for us as NASA plans for a space station, a moon base, and while a manned trip to the planet Mars begins to take shape. Dr. Goddard's prediction that men would ultimately conquer their inertia and become universal explorers has made good progress in only a short time. Dr. Goddard was certainly a "Spokesman" for God in his time.

What gave him such a vision? What gave him such faith in the orderliness of the universe? Faith comes from God's word, the Bible. This scientist no doubt trusted in the Holy Word more than most of us and received much of his inspiration from reading and studying the Bible.

Esther Goddard was faithful in deciphering and transcribing his notes, especially as his illness of tuberculous and throat cancer progressed. In the spring of 1943 she described a note he made in his diary beside a sketch of an iris blossom after attending chapel services at the Annapolis Academy: "Note from sermon: sun-crowned man."

She wondered if he had copied it down because he felt himself to be a sun crowned man. As for her, she definitely felt that her dear husband was indeed a "sun-crowned man." The world of science and technology would be much more advanced today if more wives would give better support to their husbands working in the busy aerospace fields. I remember how hard the families suffered during the early years of the United States missile and space program. Most of us were working long hours in work clothed in secrecy. We had to leave our wives out of these activities in the interest of our country's defense planning. So, needless to say, the families felt the strain. On the following page, I want to share with you the letter that Dr. Kurt Debus wrote to my wife for her contribution made during those times. Dr. Debus was the Director of the Army Ballistic Missile Agency's Missile Firing Laboratory, later the Launch Operations Directorate, and then was the first Director of the John F. Kennedy Space Center. We will have more to say about Dr. Debus.

MISSILE FIRING LABORATORY
DEVELOPMENT OPERATIONS DIVISION
ARMY BALLISTIC MISSILE AGENCY
REDSTONE ARSENAL, ALABAMA

4 February 1958

Dear Mrs. Childers:

At the time I extended my congratulations to the active members of the Missile Firing Laboratory for the part they performed in the final preparation and launch of the first U. S. satellite, I was again conscious of the many intangible but important contributions made by the wives of those who are required to work long and strange hours in our business.

It is fitting, therefore, that you also be recognized for your contributions to ABMA and the Missile Firing Laboratory, and I wish to take this method of expressing my sincere appreciation.

I hope that you as a "silent but very important" member of our team join in our pride of accomplishment, not only at this time, but always.

Sincerely yours,

KURT H. DEBUS
Director, Missile Firing Lab

A Copy Of Dr. Debus' Letter To My Wife,
Dated February 4, 1958

Konstantin Edwardovich Tsiolkovsky

As Dr. Goddard allowed his manuscript on *"A Method of Reaching Extreme Altitudes,"* to be published in the Smithsonian Miscellaneous Collections, Volume 71, Number 2, 1919, he did not know that a pioneer physicist had preceded him - a school teacher in the small Russian town of Kaluga,named Konstantin Edwardovich Tsiolkovsky. The May 1959 issue of *Astronautics* magazine (now out of circulation) published a short autobiography of Tsiolkovsky:

> I was born on the 5th of September, 1857, in Ijevski Village of Riazan Province, Russia. My parents were poor, my father an impractical inventor and philosopher. Mother, as father used to say, had a spark of hidden talent; among her relatives were many gifted people. I was eight or nine years old when she showed me a small toy balloon made of collodium and filled with hydrogen which aroused my interest.
>
> At the age of 10, my hearing was impaired by the aftereffects of scarlet fever. Being almost deaf made me a victim of ridicule to the rest of the boys in the neighborhood. This handicap estranged me from people and prompted me to read, concentrate, and dream to keep from boredom. Hurt pride sought satisfaction. There was a desire to do something big, heroic, to be rewarded, and at 11, I tried to express myself by composing silly verses. At 14, I gained from a physics textbook some theoretical understanding of aerostatics, and tried to fill a sack made of tissue paper with hydrogen. The experiment failed.
>
> At that time I was also interested in mechanical flight by means of flapping wings. Among other things, I made a small model of a carriage which moved in all directions, with and against the wind generated by bellows. There was another model of a carriage powered by a steam turbine. Later, I decided to build a larger

carriage for myself, suitable for travel, spending lunch money on lumber, screws, nails, etc. The project failed due to lack of patience and money and the final realization that the thing was impractical.

At 15 or 16 my knowledge of elementary mathematics gave me a better understanding of physics. Above all, I was interested in static balloons and already had enough data to solve the following problem: To find the dimension of a balloon, constructed of metal of a certain thickness, that would lift a given number of people. It was clear to me that the thickness of the skin will increase indefinitely as the balloon grows in size. Since then, the idea of a metallic aerostat stuck in my mind. It tired me out, and then for months I was occupied with something else, but I invariably came back to it.

Systematically, I studied very little and read only what could help me solve problems that interested me at the moment. For instance, the theory of centrifugal force interested me since I thought it could be used as a means toward the end of flight in cosmic space.There was a moment when it appeared to me that I had solved this problem (at l6). I was so excited that I could not sleep the whole night, and instead spent it wandering through the streets of Moscow and thinking about the great consequences of my discovery. Toward morning, I was convinced of the fallacy of my invention. I still remember that night, and even now, 50 years later, I sometimes dream about rising in my machine toward the stars and feeling the same exaltation.

My tribute to perpetual motion, thank God, was small. The confusion lasted only several hours due to an incorrect understanding of magnetism.

The thought of communicating with cosmic space never left me. It prompted me to study higher mathematics. Later, in 1895, I cautiously expressed my conjectures in "Dreams about Earth and Skies: and, still later (1898), in "Investigations of Cosmic Space by Reactive Machines," published in Scientific Survey, 1903.

> Astronomy attracted me because I always thought that not only the earth, but the whole universe is the heritage of mankind. The articles, "On the Moon", "Gravitation as a Source of Cosmic Energy," and others prove my constant interest in astronomy. (10)

I end my quotation from this short autobiography of Tsiolkovsky with the observation that he believed in God and that he had great insight into God's plan for himself and his fellow man when he expressed the thought "that not only the earth, but the whole universe is the heritage of mankind."

Hermann Oberth

Three years after Dr. Goddard wrote his first significant book, *A Method of Reaching Extreme Altitudes,* published by the Smithsonian Institution in 1920, a German mathematician by the name of **Hermann Oberth** published a similar book, *By Rocket into Inter-planetary Space.* A third edition was called *Ways to Space Travel.* Both Goddard and Oberth followed the same research path and practical experiments in order to put their theories into practice. Germany proclaimed Professor Oberth the world's leading expert on astronautics during the 1930s when the German Rocket Society was formed, and he became involved with the adventuresome task of building and launching liquid-fuel rockets. At that time a young member of the German Rocket Society became a student of Oberth's and gave the Professor much credit for his having followed rocketry as a profession. His name was Wernher von Braun.

I was present at Huntsville, Alabama when Professor Oberth came to visit the Redstone Arsenal where we were building the Redstone rocket. He was a most impressive man with a very dignified air about himself. I met him at the time and felt deeply humble in his presence. The reason I refer to him as Professor is because I heard Dr. Von Braun refer to him by that title. I actually heard him call Oberth, "Professor," as he introduced him around the labs. During the Apollo Program at Kennedy Space Center, I recall that Professor Oberth visited us for brief times as a consultant or as an associate of Dr. Von Braun.

Hermann Oberth was born in the summer of 1894 in the town of Hermannstadt, Hungary. His father was the respected Dr. Julius Oberth who ministered to the whole town and surrounding countryside.

When infant Oberth was baptized, the whole town was interested and predicted great things of this young infant boy. From the beginning his father wanted him to become a doctor and follow in his footsteps, but early on, young Hermann put his studies and thoughts into the sciences, mathematics, and in the sky where he thought of traveling in space. As a boy in school, he experimented with the forces of gravity and other things of nature that interested him more than the study of medicine. He read the book *From The Earth To The Moon,* by Jules Verne while still a boy. He read it through six times, drew pictures of the spaceship, Columbiad, and copied down Verne's calculations. Every day he searched for ideas to apply to space, always wondering what could be used for fuel, and how the human body would withstand the rigors of the space environment.

Hermann was caught up in World War I and served as a medic in the trenches near the Carpathian Mountains where his unit had the task of keeping the Czar's Cossacks in their own country. His soldier duties did not keep him from dreaming, experimenting, and drawing his sketches of a space vehicle and the inner configurations which he envisioned for a successful vehicle. After the war, he first enrolled at the University of Munich, then moved to Gottengen University in the hopes that he could find lodging for his family, but to no avail. He then transferred to Heidelberg University. In order to graduate, he had to write a thesis. He decided to use his book, *Die Rakete zu den Planetenrdumen* (By Rocket to Interplanetary Space) as his thesis to graduate. While waiting for its acceptance, he wrote to Dr. Robert H. Goddard, head of the Physics Department at Clark University in Massachusetts, but failed to get an answer. He also failed to convince Gottengen University to accept his thesis, so he tried several times to get a publisher to publish his papers, to no avail. His wife, Tilly, advised him to publish it himself and offered her savings to help. His book was published as a "paperback," and sold very successfully.

After this success, he was offered membership in the German Rocket Society where he met the young German student, Wernher von Braun, who took much instruction and information from Oberth as the society began experimenting with liquid fuel rockets while they both dreamed of space travel vehicles and did preliminary design and hardware development for such future work. Later in the war, Oberth was recruited to help at the German V-2 launch site at Peenemunde where he saw his work manifested in a vehicle of destruction instead of in a vehicle built for space travel. His associates there later came to America to help the U.S. Army develop missile, satellite and space vehicles. Wernher von Braun sent for his former professor when the team came to Huntsville, Alabama to work. While there the Iowa Wesley College presented him with an honorary degree, finally making him at last, "Dr. Hermann Oberth." The West German Government presented him its highest award to a scientist, the "Great Cross of Merit." At last the world began to show this gentle scientist the respect and honor that he deserved. In order to give us some insight into the life, character and vision of this great rocket scientist, I quote from the book by Helen B. Walters called, *Hermann Oberth: Father of Space Travel*, published by The Macmillan Company, New York. This is the Introduction written by Professor Oberth:

> I am very grateful that this book is being published in the United States, not so much because it is a personal honor, but because I hope it will be useful for all humanity. The discoveries, inventions, and developments of today are not the work of one nation or one man, but of all countries and all mankind throughout the ages. By concord and agreement, progress is made and science and technology are advanced; by discord, man's efforts are scattered and the products of his work are destroyed.
>
> It is fortunate that the enemies of yesterday have been reconciled, as far as it is possible, for modern rocketry and space flight provide a common task for all mankind. Many nations and innumerable people have contributed to the Space Age. I myself was able to offer the mathematical theory of rocket performance, the theoretical proof of the feasibility of space flight and a discussion of the problem of weightlessness, as well as suggesting and developing such devices as liquid

oxygen-alcohol rockets, multistage rockets, regenera-ive cooling, veil cooling, automatic steering by inertial devices, gyroscopes and servo motors, and the space mirror. All these were outlined in my first book, *Die Rakete Zu Den Planetenrdumen* (By Rocket to Interplanetary Space), first published in 1923.

Other men in other nations have contributed their share to the task of making the theory of space flight a reality. Such things as manned satellites and liquid oxygen-hydrogen rockets, invented and developed by scientists of different countries, today help to show that mankind is becoming a great family, in which each member depends on the others.

There is much to be learned from the world events that occurred during my lifetime. One of the reasons why Germany lost World War II was that too many people, both civilians and military personnel, were interested only in their own advancement. They forgot that no one benefits when personal ambition is realized at the expense of one's country, and that it is sometimes necessary to deny oneself a profit or an honor when it may be harmful to all.

At that time there were also too many learned men who did not understand the importance of long-range rockets and atomic bombs. It is necessary for a modern scientist to study so much in his own specialized field that like the goose stuffed with food cries, "God save me from more!" He is often hostile to the new ideas of those in other branches of science.

This state of mind hinders the work of discoverers and inventors and cannot be changed by greater specialization, which, though it may make one's own work easier, also restricts the mind. It is my feeling that better understanding would be possible if scientific material were concentrated into good manuals and further publication were limited by the practice of writing a scientific report only when it offers original material or improves existing material.

There is something else I would like the reader to

notice while reading this biography, I am sure I could have lived more comfortably, especially during my youth, had I worked in the conventional way -- had I become a doctor as my parents wished, worked only for financial gain, and done what people wanted of me. However, I would not have been happy! I believe that if one feels called to a certain profession, he should try to follow it without allowing such superficial considerations as position, money, and honors to affect his decision. A man should do the work he enjoys; however, once he has chosen his vocation he should learn it thoroughly and also should not shrink away from difficulty and danger. Henry Ford, in his biography, said that no one should attempt to invent an engine he could not build with his own hands. In 1930 I was given the chance to build a rocket in Berlin, but I was unable to deal with all the practical problems of construction and the rocket was unsuccessful. Afterward, I learned the trades of the locksmith and mechanic. Had I known them earlier, my work in Berlin might have been a success.

Finally, although I do not believe, as men once did, that our earth is a disk surrounded by a bell-shaped sky, over which God and His Angels watch to see if we are good, I have come to believe that life is not explainable in terms of matter alone. Man is not merely a machine; he possesses an immortal soul that is the source of the phenomena of life: experience, will, love, and art.

The belief that each man must do what he is able has helped to carry me through many difficult times: I have always felt responsible to the Lord for the use I made of the abilities He gave me. (Reprinted with permission of Simon & Schuster Books for Young Readers, an Imprint of Simon & Schuster Children's Publishing Division from HERMANN OBERTH, FATHER OF SPACE TRAVEL by Helen B. Walters. Copyright, 1962 Helen B. Walters). (11)

Here once again a great rocket scientist gives God the credit for what he was, and for what he was able to accomplish in the domain of space and for the future of space travel.

On January 22, 1990, the Florida TODAY newspaper reported on an interview that had been held with Professor Oberth at the John F. Kennedy Space Center in which he was proclaimed as the father of the planned International Space Station. I quote:

> "....Checking our interview notes, we see that we asked Oberth how it felt to be the expectant father of the Space Station.
>
> "Most satisfying", he said. "It is the culminating event of my career, personally and within the context of the space program".
>
> He hoped the Space Station would foster an era when people will be more friendly, when they will work more by reason than by some irrational feelings, when they will work together not only for science but for peace.
>
> On the technical side, what would give him most satisfaction would be to see the space station made and then used to bring home some improvements for Earth out from space." (12)

Today, June 17, 1997, I received a copy of the Kennedy Space Center's "Spaceport News," which tells of the arrival of the first element of the space station which Professor Oberth was talking about in the above statement. Only six and a half years have passed since he prophetically uttered those thoughts about the Space Station to the Florida TODAY newspaper columnist.... {In that vain, aero-and astro-technology advanced unbelievingly fast when one remembers that only sixty-one (61) years lapsed from Kitty Hawk, N.C. to Tranquillity Base on the moon, 1908-1969; (the author's comment)}

The Professor's hope was that it would be a truly international outpost. He earlier mentioned that the world is becoming a great family where each member country depends upon the other.

The Professor was truly a "spokesman for God," as evidenced by his statements of faith and by his almost "holy calling" as one of the "fathers of space flight."

Wernher von Braun

These three "Fathers of Space Travel" did not live long enough to put all of their plans for space travel into effect, but following in their footsteps was that young German boy I mentioned earlier by the name of Wernher von Braun, who became interested in the work of Professor Oberth.

Wernher was born in 1912 into a family who had traditionally served in the government. His brother, Sigismund, had served as a representative to the United Nations. Their father, Baron Magnus von Braun served in his local and central government for a time. The Baron supported his three sons in whatever they chose as a profession. Wernher's mother was well educated and was an outstanding astronomer and natural scientist. She had almost a religious outlook toward the orderliness and beauty of nature. Wernher was eager to learn from her and by the time he was ten, he knew most of the stars and their constellations and he looked at them as they moved across the sky.

Baroness von Braun had a firm hand on her boys and believed in discipline for their good. Both parents believed in giving their sons the benefits of a good religious education as well.

Wernher continued to read about his favorite subject and came across Professor Oberth's book, *By Rocket Into Interplanetary Space,* which gave him further inspiration. He sought out the Professor and became a student and associate of his as they worked together in perfecting liquid-fueled rockets for the German Rocket Society and later the German Army. At the rocket proving ground and launch site called "Peenemunde," Wernher led his team in developing the V-2 rocket which rained death on England and Belgium during World War II. All the time he worked on military rockets, he dreamed of putting those rockets to use in the conquest of space. He wondered where he could get support for such a dream. At the close

DR. WERNHER VON BRAUN, First Director,
George C. Marshall Space Flight Center (NASA Photo)

of the War, Wernher and a group of 129 scientists, engineers, technicians and administrators from Peenemunde elected to surrender to the American forces and come to America where they could possibly pursue their dreams. This team of German rocket experts brought about 100 of the V-2 rockets with them to Fort Bliss, Texas where they did development work there and at White Sands, New Mexico for the U.S. Army. Later they moved to Huntsville, Alabama where I joined them as a young Electronic Engineer. I worked on the Redstone Missile instrumentation equipment which gathered the health and performance data of the missiles and transmitted the data to the ground telemetry station via radio. We static tested the missiles at Huntsville and then transported them to Cape Canaveral, Florida where we launched them down the Long Range Proving Ground over the Atlantic Ocean.

Our team, under the direction of Dr. Von Braun and Dr. Kurt Debus, launched the first free world satellite on January 31, 1958, and put the first American into space (Alan Shepard) on May 5, 1961. Through these activities, Dr. Von Braun began to see his dream of traveling in space come true. The creation of the George C. Marshall Space Flight Center at Huntsville, Alabama, at last allowed him to devote his total effort to that dream. He managed the development of the mammoth Saturn rockets for manned space flight to the moon and return. The Saturn V launch vehicle was derived primarily from Von Braun's designs and ideas. We all know that the moon landing was achieved on July 20, 1969. In 1972, Von Braun resigned from the National Aeronautics and Space Administration and became president of the non-profit National Space Institute, formed to promote space travel. He had always dreamed of setting foot on the moon himself, but this dream was never realized because cancer ended his life on June 16, 1977.

It was during one of the Saturn launches that I last saw this great man. He was visiting at the Launch Control Center and we rode together up to the third floor on the elevator. He recognized me and greeted me with his usual graciousness. It was really great to see him if only to be in his presence for a few minutes.

In just being around Dr. von Braun, one could sense his great faith in his men and in his mission in life. All of us on his team looked to him for inspiration and guidance. Even the U.S. Congress looked to him for recommendations on the national space goals and scientific endeavors. In an age of charismatic leaders he certainly exhibited the popular attributes of one - - his ability to communicate confidence and strength, his ability to enlist and enliven our loyalties, and to uplift, encourage, and empower people to dare and to care.

When we look for a definition of the word, charisma, we find that Apostle Paul adopted the word and used it to describe the style of life of a person who had received the grace of Christ. I would like to share with the reader some of the research into the life and faith of Dr. Von Braun and his measure of this "Grace of Christ."

The following interview of Dr. Von Braun by Evangelist C.M. Ward is provided with permission from the Evangelism Literature for America (ELA) and gives us a glimpse into the faith of this great scientist:

We talked about space and faith.

The interview with Dr. Wernher von Braun took place at his conference table in his offices at the George C. Marshall Space Flight Center, National Aeronautics and Space Administration, Huntsville, Alabama, on May 9, 1966. He is quick to appreciate Biblical comparisons. When I suggested that the Tower of Babel may have been the first launching pad, he responded with:

> "I think there is another similarity. I understand that the project was never finished because the various builders did not understand each other's language. We have that problem in modern science. There are so many specialists involved that it is becoming increasingly difficult to intelligently communicate with each other."

Dr. Von Braun's moral mandate is simple:

> "God had built man with curiosity. God expects man to use this gift. I do not think the process should ever stop. Now we have the tools available to explore space. I believe this: if it were not the Creator's intent for us to explore celestial bodies, He would not have permitted us to have acquired the tools. Something else is apparent. God has not placed any visible obstacles in our way. I believe we have His permission and His blessing."

The faith of this world-renouned scientist is the faith of an open mind moving freely through the impetus of an unshakable conviction that "God....is, and that He is a rewarder of them that diligently seek Him" (Hebrews 11:6). This faith says every day in the midst of the most complex laboratories the earth has ever known, "and it doth not yet appear what we shall be" (1 John 3:2).

The Doctor returns to this point repeatedly:

> "There are other dimensions involved than just time and space. Prayer cannot yet be measured by technicians. Nevertheless, prayer is a factor."

We talked about his days at Peenemunde under Hitler, and about the closing hours of the struggle when he and his staff of rocket scientists were under tremendous pressure. He said:

> "We knew the potential we possessed. We realized that as far as Germany was concerned there would be no further opportunity to develop this potential for a long time to come. So we were confronted with the question - Into whose hands would we place this? We had wearied of dictatorship in any form or fashion. The idea of winding up under Stalin after all those years under Hitler disturbed us. We sat together and discussed it.

We were unanimous. We wished to join the Americans. Personally, I was strongly convinced that since we had this treasure it was our duty to put it in the right hands."

Dr. Von Braun believes resolutely that Providence directed the implementation of this decision. And he says:

"Periods of perplexity come. I ask, 'Am I or am I not on the right track?' It is during these moments of questionings, especially that I am strengthened by asking for and receiving help from God."

He put it very simply and directly when he said:

"My need for guidance had grown through the years. I think the reason is that in work such as mine you cannot get a lot of handbook advice as to what is right and what is wrong."

What discipline and rule of thumb does he accept?

It has not changed since his father inculcated Christian principles into him as a lad in their ancestral home, Silesia. He believes in a final audit - a Last Judgment. These are his words to me:

"Life is a gift. I am given talent and ability in some degree. I am expected by God to do something with these. I am accountable."

He has said on numerous occasions:

"I think science has a real surprise for the skeptics. Science, for instance, tells us that nothing in nature, not even the tiniest particle, can disappear without a trace. Think about that for a moment. Once you do, your thoughts about life will never be the same.

Science has found that nothing can disappear without a trace. Nature does not know extinction. All it knows is transformation.

Benjamin Franklin, a scientist, put it well. I believe that the soul of Man is immortal and will be treated with justice in another life respecting its conduct in this. Now if God applies this fundamental principle to the most minute and insignificant parts of His universe, doesn't it make sense to assume that He applies it also to the masterpiece of His creation - the human soul? I think it does. And everything science has taught me - continues to teach me - strengthens my belief in the continuity of our spiritual existence after death. Nothing disappears without a trace."

Dr. Von Braun believes that Bible faith makes a difference in this life. He looks back, and what he sees is razor sharp in unanswerable argument. He saw Hitler put together a very brilliant and coordinated economy. For years now as an American citizen, he has seen this economy grow and reach toward the ends of the earth. What is the difference? He replies:

"I'm absolutely convinced that Hitler was not a God-fearing man. He had no scruples whatsoever. After he had reached the top there were no checks or balances left to govern him.There was no one to whom he was accountable. This evil virus infected the genius of organization and technocracy. It was a rodent that ate away the strength of the nation. Faith is the difference."

What is out there?

“Sometimes one thinks that Lazarus in his "space suit" is almost the progenitor of today's headline stories of astronauts sent into space and returning. Lazarus came out of his capsule by the command of Jesus, and stood wrapped in clothing the size and shape of space wrappings today. The Master said, "Loose him, and let him go" (John 11:44). Lazarus had been into space and back. Can we find in life what has been believed we can find only in death?"

Dr. Von Braun believes this generation will no longer have to look through "the dirty basement window of our atmosphere." We are going to have a better look:

"By setting up observation stations in space we will have a platform to learn far more about the universe. Through a closer look at creation we ought to gain a better knowledge of the Creator. A greater sense of man's responsibility to God will come into focus."

Dr. Von Braun said in a prepared speech given in Washington, D.C., to a forum sponsored by the international Christian Leadership Conference, February 17, 1966:

"Because of the wonderful things it has done for society, we are tempted to place too high value on science. We should remember that science exists only because there are people, and its concepts exist only in the minds of men. Behind these concepts lies the reality which is being revealed to us - but only by the grace of God."

Our conversation veered to personal responsibility. Are there moments of decision in life?

He answered:

"You know what I have been through in Germany. When I came to this country I had time during my early residence to do a lot of stock-taking. I asked, "What went wrong? What should be changed henceforth? How much was I personally involved? War is a cataract that suddenly widens and gathers you in. Yes, I have had those moments."

Does Dr. Von Braun wear his science clothes six days a week and then put them off and reach for his religious clothes for Sunday? No, not at all! His life is compartmentalized. This is how he views his life:

> "I think science is basically an attempt to better understand creation. Since I cannot contemplate a creation without a Creator or a Divine intent I believe science has the same motivation as religion - communion with the Creator."

A strange impact has branded the Doctor. Let him describe it!

> "When we were brought to Texas I saw a difference immediately in church life. I had seen the huge cathedrals of Europe. They represented fabulous architectural and art values. However most of them were empty, especially during the war years when worship was discouraged. Now in America we saw all kinds of churches, many of them in temporary buildings, and thriving second-hand busses gathering congregations. I was impressed. I thought, 'Here is a growing, aggressive church and not a dignified, half-dead institution. Here is spiritual life."

Dr. Von Braun's conclusions are profound.

He says:

> "I know of no scientific discovery that would necessitate a change of relationship between God and man. Is it not a real miracle that into whatever circumstances fate places you, there is always that sensitive little needle before your inner eyes that is called conscience? It tells you unequivocally what is right and what is wrong."

He says:

"A scientist sees God's handwriting on the wall wherever and at whatever he looks. He also asks himself the question, 'Why do I want to know this?" That, of course, is almost tantamount to the question, "Who am I?'"

Now it is time for us to ask ourselves the same question, "Who are we?" David pondered the same question centuries ago:

"When I consider thy heavens, the work of thy fingers, the moon and the stars, which thou hast ordained; what is man that thou art mindful of him? And the son of man, that thou visitest him." For thou hast made him a little lower than the angels, and hast crowned him with glory and honour. Thou madest him to have dominion over the works of thy hands; thou hast put all things under his feet" (Psalm 8:3-6).

At the conclusion of this interview between Evangelist C.M. Ward and Dr. Wernher von Braun, Dr. Von Braun declared:

"Without the recognition of a divine Intent, the developments in the animate and the spiritual worlds simply cannot be understood."

In the 1950's I began to gather resource material for this book on "Faith In Space." Among this material were some magazine articles and quotations from books I read which further establishes Dr. Von Braun's faith in God, and in all God's creation. The following quote is taken from the October 1959 *Together Magazine*, from an article entitled, "Missiles and Civilization" by Wernher von Braun:

"One by-product of the ongoing missile race between Russia and the U.S. is an increasing hesitancy among many conscientious Americans about the rightness of venturing into space. A feeling appears to be growing that perhaps man has no business trying to extend his domain beyond earthly limitations - that space exploration is an affront to God."

Dr. Von Braun said that technology is the key to liberating man from his bondage to physical labor so that he may devote himself to the development of his mind and spirit, and that space travel is the ultimate development in technology. He stressed that once having entered upon the road of technology, no civilization can turn back. The old pattern is destroyed in the development of the new. He was concerned about the ethical standards of scientists and government leaders. He said:

"If the world's ethical standards fail to rise with the advances achieved by technology, we will perish. Technology and ethics must go hand in hand. While technology controls the forces of nature around us, ethics controls the forces of nature within us. But while technology is a mere 150 years old, the problems of ethics have occupied the minds of the world's greatest thinkers for centuries. I think it is a fair assumption that the Ten Commandments are entirely adequate, without amendments, to cope with the problems of using technology, both present and future.

What is lacking is not an ethical standard but sufficiently tight, day-to-day application of the standard we have. This lack stems from our separation of science and religion, a separation which manifests itself in the belief that scientific enlightenment and religious belief are incompatible.

For all the scientific enlightenment of the last two centuries, we know of more mysteries in nature today than when the technology revolution began. Science has in no way done away with God; it has only broadened the frontiers along which we can see His wonderful works. Science has not taken God away from man; man has taken God away from science.

As technology advances, it will free him to devote more of his time to thinking and dreaming; it will enable him to raise civilization to new levels.

When space flight frees man from the final chains of gravity, it will open to him a new and wonderful experience of God's truth. It will open to him the gateway to heaven." (13)

Erik Bergaust, in his book, WERNHER von BRAUN, page 108, revealed the following in Dr. Von Braun's own words after taking a cross-country glider trip to try for his "Silver C" Certificate: "Of course, God is everywhere. And you see Him all around you all the time. But when I am up there - alone - feeling my way through the clear skies and listening to the soft rushing of the winds over the surfaces of my glider, then, somehow, He comes on particularly strong. The closeness becomes very real."

I will close this chapter with one more example of Wernher von Braun's faith in God as taken from the little book entitled: *QUEST*, Tyndale House Publishers, Inc. Wheaton, IL, by Steve Carman, an aerospace engineer like myself, who worked for North American Aviation in Downey, California.

On page 70 of his book Steve made a statement which further convinced me of Dr. Von Braun's deep faith in God.

He overheard this interchange between Dr. Von Braun and a reporter as they were being conducted through the Apollo Command Module high-bay clean room. Mr. Carman gave me permission to quote:

"As a Christian, Dr. Von Braun," the reporter had asked, "You must have second thoughts about traveling into space to look for life, don't you? For if we find life out there, won't it mean that there is no God?"

"On the contrary..." The eminent scientist had responded in a way that had caused me to think very deeply."...I am now more interested than ever before in a search for life in the universe, for if we find another civilization in space, I believe it would be the greatest proof of God's existence man has ever known. You see, I believe that God created the universe, including man. And if there are other beings in this universe, they exist only because God created them...and they would then know the same God that I know."

Dr. Von Braun had talked about the possibility of a world where there was no "sin," a planet where men lived in harmony with God, "as intended in His original creation."(14)

In Erik Bergaust's book, WERNHER VON BRAUN, Cob and Dunlop, Publishers, p. 117, Dr. Von Braun further witnessed to the readers when Mr. Bergaust asked him, "Haven't there been times in your life when you felt the need to pray -- kind of particularly strong?" He answered: "I certainly prayed a lot before and during the crucial Apollo flights, and I also prayed during the last days in Germany -- when things collapsed all around me. Indeed, during those hours of decision, when we decided to surrender to the Americans, my anxiety was at the bursting point. I prayed then that our surrender would be accepted in good faith."

History certainly provided answers to Dr. Von Braun's prayers in most matters concerning his surrender to American soldiers and to all the concentrated work the Von Braun team accomplished for the United States missile and space program. One can certainly see God's hand in all phases of his team's work as they pursued their life-long careers in space flight. Praise God for placing such God-given talent into our hands. May America never forget what we owe Dr. Wernher von Braun, who by choice and devoted service led our country into the Space Age. Truly , he was a *Spokesman For God.*

Chapter 4 - ENGINEERS BELIEVE

How do Engineers get involved in space travel? They take the dreams, the ideas and possibilities projected by the space scientists and others and turn them into reality just as we see accomplished in the following "Design Reference Mission" developed for the Apollo Program:

> The mission begins with the roll-out of the space vehicle from the Vertical Assembly Building approximately 20 days prior to lift off. Launch occurs on September 17, 1969 with a 72 degree azimuth at 07:49 A.M. A near-circular parking orbit is attained with an altitude of approximately 100 nautical miles. Translunar injection (TLI) begins over the Pacific Ocean during the second orbit, and results in a free-return circumlunar trajectory.The translunar coast plane is inclined 32.6 degrees to the earth's equator and 28.7 degrees to the moon's orbital plane. The transportation maneuver begins 15 minutes after injection and results in the completion of hard docking and S4-B jettison 30 minutes later. The time from beginning of coast to pericynthion is 61 hours, during which time the crew follows an assumed work and rest schedule. The orbital coast period after docking includes two passes over the landing site before transearth injection (TEI), for navigation and checkout purposes. TEI is initiated on the 22nd CSM orbit, and includes a 1.8 degree plane change during the burn. During the transearth coast, which lasts 89.1 hours, three midcourse corrections are made. These are assumed to occur at 20 and 65 hours after TEI and at two hours prior to entry. Again, no corresponding perturbations to the trajectory are illustrated. Entry begins at 380,759 feet altitude with a range-to-go to the Hawaii landing site of 1570 nautical miles. The drogue chutes are deployed 10 minutes and 28 seconds later at an

altitude of 24,000 feet. The drogue chutes are jettisoned 48 seconds later, followed by deployment of the main chutes at about 10,000 feet altitude. Touchdown occurs near Hawaii at 152 degrees west/17 degrees north 198 hours, 33 minutes, and 7.6 seconds (approximately 8 days and 6 hours) after earth launch.

Without dedicated engineers and technicians we could not have progressed very far into the space age. How do they feel about God and about man's dominion in space?

First let us look into the life of one of the greatest engineers of our time (the person that President Eisenhower put in charge of launching our first satellite), Major General J. Bruce Medaris. General Medaris became an ordnance officer as he progressed in his military career. An important bonus of being an ordnance officer was the chance to acquire a master's degree in engineering at Massachusetts Institute of Technology. He served with distinction during World War II and was awarded the Distinguished Service Medal.

My first impression of General Medaris was that of an arrogant, self-centered Army General. I had seen him on many occasions when he visited the Army Ballistic Missile Agency's launch complexes 5, 6, and 26, at Cape Canaveral. I had observed him pushing into the blockhouse with his swagar-stick in hand, seemingly looking down at us engineers and technicians as we prepared and launched the Redstone and Jupiter missiles. The only time I saw him lay his swagar-stick and hat down was when the Secretary of Defense, Charles E. Wilson, came into the blockhouse with us. He immediately handed his hat and swagar-stick to his Aide, and I heard him address Mr. Wilson as "Mister Secretary" as they conversed about the up-coming launch. How did this self-confident Army General, commander of one of the most important Army posts, and the man who led the United States' effort to launch the first free world satellite, come to be a humble servant of God?

Gordon Harris, in the life story of General Medaris,

MAJOR GENERAL JOHN BRUCE MEDARIS, March 1958
"He pushed the button on the Space Age" (ABMA Photo}

A New Command, (Published by Logos, International) shows a different person than one perceived upon first meeting. In many instances Gordon documented the fact that General Medaris prayed for guidance when a big decision was before him.

On the occasion of his preparing to come to the Cape to launch Explorer One, his wife Ginny was folding shirts for him and in their conversation, he said: "This is our biggest challenge. We've waited a long while for recognition and now we must make good on our promises. I know that Wernher and his people are pushing themselves to the limit but in all honesty, Ginny, this is such a complex operation that the failure of one little part could defeat us. <u>I'm praying for help</u>."

General Medaris and Wernher von Braun were considered as the "Twin Fathers of our space program" by many of us who knew all the early history of the space program. Medaris was not only instrumental in orbiting the first satellite, but he championed all our efforts in planning to put man into space. He was put in charge of the Army Ordnance Missile Command which included the Jet Propulsion Laboratory, White Sands Missile Range, Redstone Arsenal, the Army Rocket and Guided Missile Agency, and test units at Point Magu, Fort Churchill, China Lake, Kwajalein Island in the mid-Pacific and an office in Paris that dealt with NATO.

When NASA began planning to launch Americans into space, the Associated press asked General Medaris and Wernher von Braun the same question, "Will the intoxication of man's first flight in space cast doubt on the existence of God?" Their answers appeared in newspapers July 8, 1959.

General Medaris answered:

> "The ten commandments have guided men and safely brought them through upheavals of the past. They should be the rock of man's decisions today and

tomorrow. I am convinced that our scientific advances were made possible by the Grace of God. Leading figures of all major faiths including Pope Pius XII saw no defiance of Divinity in our initial ventures into space. The door would have been closed in our faces if what we are attempting contravenes God's will. Our first steps have reemphasized the beauty and order of creation and verified natural laws which govern all existence. As he ponders these things man must come to believe anew, as once he did before false intellectualism, bigotry and superstition turned him away from truth."

How did this arrogant, self-centered commander of one of America's most important military posts become a devoted servant of God? The answer lies in his own words as presented in the September, 1968 issue of *Guideposts* magazine:

"In 1960 I decided to retire from the Army, and in 1962 I formed my own company in the field of space technology. My wife and I built a house near Orlando, Florida. We were looking forward to the best years of our life together when suddenly in 1964, like lightning out of a black cloud, cancer struck me once again.

This time, I was told it was bone cancer. The X-rays showed that it was scattered throughout the pelvic area. The doctor said he could help relieve the pain, but that there was nothing that could be done about the cancer itself. I asked him to be honest with me about the future. He gave me between a year and a year and a half to live. In the Army you learn to accept things stoically. I knew that the members of the little church we had joined were praying for me and I appreciated it. But I didn't put too much stock in such prayers. I had seen those X-rays.

Then, in rapid succession, two things happened. First a visitor stopped by our house and left a book called *"God Is Able"*. I never got beyond the first chapter, but I didn't have to. Reading those words, I was struck by the overwhelming conviction that God not only is, but that He can do anything He chooses to do, that He can and does intervene in human life. The result of this sudden, almost blinding, flash of insight was that I stopped being doubtful about the power of prayer and began to ask urgently for help.

Next we were invited to a friend's house to meet a housewife who, we were told, had a special gift of healing. I had always looked askance at anything in the realm of spiritual healing. But somehow I knew that I had to go. I not only went, I found myself asking for help. And when this housewife, who seemed to be the most normal and good-humored of people, put her hands on my head and said a short prayer, I felt as if 2,000 volts of electricity shot through me, leaving me stunned and weak and somehow *different.*

As Francis Thompson also says in "The Hound of Heaven":

"*Yet ever and anon a trumpet sounds*
From the hid battlements of Eternity....."

That night, such a trumpet sounded for me. After all the years of running and hiding, of doubting and questioning, I let go of the last tattered shreds of ego and surrendered myself, body, mind, and soul to Jesus Christ. "My life is Yours." I said to Him. "Take what's left of it and use it in any way You want."

After that, my next medical examination showed no further growth of the cancer. I was no better, but neither was I worse. The doctors could not understand this, since the rate of progression in bone cancer of this type has been quite well established. Subsequent X-rays showed that the cancer was regressing. Today - more than two years after the date the disease should have killed me - no lesions remain except for a trace of the largest one that was there originally.

What does this mean? To me it means that God's unbelievably patient and tolerant and understanding love will pursue a man - even a man full of the sin of self as I was until it catches up with him.

For years I thought that my life had been spared, over and over again, so that I could play a part - an enormously fulfilling and ego-satisfying part - in placing a metal sphere into orbit at 17,500 miles per hour around the earth. But perhaps the real reason was so that I could tell this story of my own personal encounter with the "Hound of Heaven." It's an encounter that has left me filled with an abiding joy.

My experience gives me the right, I think, to say to any troubled, uncertain, doubting person who may be reading this: Wait. It happened to me. It can happen to you, too". (15)

General Medaris did not stop at this point in his Christian walk. On June 19, 1969, this strong impressive Major General took another step toward being a servant of God -- he was ordained a Priest in the Episcopal Church. The Florida TODAY newspaper of January 3l, 1988 chose Father J. Bruce Medaris to interview on the 30th anniversary of the Explorer One launching. His account of the launch preparation, the actual launch, and the pride that America felt was gratifying to recall. I personally feel like he, that God did single him out for this great task of "pushing the button on the Space Age" and for the task of taking the message of Christ to his fellow man.

I want to share with the reader one other instance of God's grace in the life of this hard-nosed General. It was at one of our annual Space Pioneer's Banquets that I heard Father Medaris give a special healing prayer for Wernher von Braun who was dying of cancer. It was a beautiful prayer and one could feel the spiritual power of this great servant of God as he gave our petition to God. In the Florida TODAY article he stated that he worked very closely with Wernher, and that he was a "wonderful guy," and that they became very good friends. After hearing that prayer, I certainly saw the truth of that statement.

Another important Engineer involved in space flight was Dr. Hugh L. Dryden, NASA's first Deputy Administrator. Dr. Dryden came to NASA after a long and distinguished career at the National Advisory Committee for Aeronautics (NACA) before it was dissolved to become a prime member of the new National Aeronautics and Space Administration (NASA), established by Congress on July 16, 1958.

Dr. Dryden was a prime mover in the areas of space flight and atmospheric research most of his life, but he found time to devote to his God as a licensed lay minister in the Methodist Church. I did not meet this great man, but being a Methodist myself, I followed his career and his leadership of NASA as the years progressed. In one of his press interviews he had this to say about science and religion:

> "Even a scientist needs religious faith," he said. "Science has made tremendous contributions to mankind. It has penetrated the mysteries of the material universe and freed the minds of men from ignorance and superstition. But science is a partial view of life, and in many respects, a narrow view."
>
> Dr. Dryden did not accept the cliche that religious people live by "faith" whereas scientific people live by "objective" knowledge. He said that some people "are led to a purely materialistic view" of the universe while others are persuaded that there are both the material and the spiritual. Either view, he emphasized, was necessarily held exclusively by the objective methods of science."

Dr. Dryden died of cancer in 1981. It was reported that only days before his death, he was busy working on an important decision involving the Apollo Program. So intense and sincere was his dedication to the nation's space program, he would not allow himself the luxury of relaxation even though he knew the end was near.

Dr. Dryden was recognized for his leadership in the

DR. HUGH L. DRYDEN, Former NASA Deputy Administrator
He Guided the United States into the Space Age.
(NASA Photo)

development of aeronautics and astronautics. In fact, he was generally regarded as the man who guided the United States into the Space Age, although he did not rest on such laurels. In the NASA History Series publication, *ON THE FRONTIER, Flight Research at Dryden 1946-1981,* by Richard P. Hallion, Page XIX, we find this account of an honor bestowed upon Dr. Dryden:

> On March 26, 1976, NASA renamed the Flight Research Center the Hugh L. Dryden Flight Research Center (DFRC), in honor of an American aerospace pioneer, a man who was fond of saying "the airplane and I grew up together," and who once remarked that "the most important tool in aeronautical research.....is the human mind."

From these accounts we already get a glimpse of this master engineer's faith in God. To further inform us of his faith in an orderly universe I want to quote from the January 1962 *TOGETHER* magazine, starting on page 14, entitled, MAN'S ETERNAL QUEST:

> "Each generation has its World of Tomorrow which looms as a challenge to its spiritual and physical courage, its intelligence, foresight, accumulated knowledge, and technical skills. Once it was the New World, unseen and unproved. Then, in turn, came other goals of geographical exploration - the polar region, the ocean depths, the limits of the atmosphere. For our generation, the World of Tomorrow lies in outer space. This supreme challenge in the material universe is second only to spiritual tests man must pass if he is to find his true place among the stars.....
>
> Who is man to dream of reaching the stars? Even when he comes to stand on dark Pluto, the outermost planet, he will find himself poised at the edge of a far vaster chasm of space that separates our solar system from others. The nearest star is 25 trillion miles away. It takes light, which on earth seems almost instantaneous in its travel, more that four years to

> make the journey. The very thought of tiny man embarking on such a trip in the puny, slow-moving space vehicles of today seems ridiculous. But already man is thinking of thermal-atomic rockets, ionic rockets, photon rockets, devices which may give him unimaginable speed and power. Within the next century, it may be possible to reach fixed-star systems without consuming the life span of the crew.
>
> One thing is certain. Man will go to the stars when the time comes. He may, of course, send his robots first, as he is doing to explore the solar system. But robots will never be enough. Machines are only machines. They cannot think as we think; they cannot write poetry, love, worship a Creator, or thrill to the wonders of a universe sprinkled with the multitude of possible worlds. Machines cannot seek God, as man has done, in earthquake and storm, in lightning and thunder, in mountain heights and trackless seas, in towering clouds and stary skies. (16)

The world owes a great deal of thanks and honor to this great engineer and scholarly leader for all that he accomplished in aerospace progress. What a spokesman for God! Also it is important to know that T. Keith Glennan would not have agreed to accept President Eisenhower's appointment as the first NASA Administrator if Hugh Dryden had not been assigned as his Deputy Administrator.

During My assignment at the Martin Marietta Corporation in Orlando, I came in contact with another great engineer whom I admired very much for his Christian witness. He was the Vice President of Martin Marietta, <u>Mr. G.T. Willey</u>. Mr. Willey was active in the Christian Business Men's Committee, made many talks and wrote the following quote for the Orlando Evening Star of January 27, 1967, where he declared his faith:

"Lent is the period of 40 weekdays from Ash Wednesday to Easter, observed in Christian churches by fasting and penitence to commemorate Jesus' fasting in the wilderness before his crucifixion. This person was God Eternal manifested in the flesh, Creator of the universe. He was the source of all origin of Life, yet He became our Man; but He is also the Alpha and Omega, the Beginning and the End. So I do not believe we should give Him consideration just during the Lenten period; but every day. He lived, He fasted, He died, and the third day He arose from the dead and today He is risen, the second person of the Trinity at the right hand of God the Father in the Heavenlies.

The event of the Resurrection is the greatest event in world history! Yes, greater than walking on the moon or the journey into space. The Resurrection of Jesus Christ from the dead places God's seal of acceptance upon the work of redemption on the Cross. It is the only true hope of the world. It changes the storm into a calm, war into peace, and darkness to light.

We read in Matthew chapter 28 that Mary Magdalene and the other Mary, as it began to dawn, walked to the sepulcher and heard these words from the guardian angel, "He is not here: He is Risen!" In these few words the mighty miracle of the resurrection of our Lord and Savior was proclaimed; and thus began the brightest dawn in history!

This event was not an accident; but the great culminating event toward which all things had converged.

The Psalmist, King David, describes it in these words; "Mercy and truth have met together, righteousness and peace have kissed each other. "The most difficult equation, since the beginning of time, has been solved. God has now bridged the gap between sinful man and the righteous and perfect God. Man lives and dies; but God died and now lives!

So I rejoice in this Lenten period not only in God's fasting; but for His complete sacrifice that I might go free. Having made a decision when I was a boy in my teens, I put my trust and my eternal security in His hands; and with the Apostle Peter, as I look back over nearly 70 years, I can say; "Being born again, not of corruptible seed, but of incorruptible, by the work of God, which liveth and abideth for ever."

This faith in Christ is the answer to the world and our country; the way out of the dilemma in which we find ourselves.

As the prophet of old wrote in the second book of Chronicles, chapter 7; "If my people, which are called by my name, shall humble themselves, shall pray, and seek my face, and turn from their wicked ways; then I will forgive their sin, and will heal their land."

Let us remember our God not only in these days of Lent; but every day and every hour."(17)

Mr. Willey was a very successful manager for the Pershing Missile Program for Martin Marietta, and a lot of responsibility rested on his shoulders since the Pershing missile was the first out-of-house contract which the Redstone Arsenal let for a ballistic missile. It was my duty to monitor the quality of work accomplished at the Orlando missile factory and to help insure acceptable items of instrumentation and telemetry systems as well as to coordinate the training of our own forces at the Cape. During this time I became better acquainted with Mr. Willey through my next-door neighbor, Mr. Joe Rinker of the Rinker Materials Corporation, who supplied the concrete for most of the Cape and Kennedy Space Center construction. Joe was also active in the Christian Business Men's Committee and vouched for Mr. Willey's sincerity and witness.

As we began to test and launch the Pershing missiles, I am sure that Mr. Willey spent a lot of time in prayer as well as in management. Most of us never entered into a countdown without first asking God to bless our efforts. And He certainly blessed our little team that did all the test firing of this first deployable solid rocket ballistic missile.

To give more insight into the belief and faith of Mr. Willey, I want to share a portion of his writing from a small Christian tract which I collected years ago:

> "It has been my privilege to have been engaged in the airplane, and more recently the missile business, for the past 40 years. One of the most important things about missiles is the firing - what we call the "count down." Usually you try to squeeze this vital phase of missile testing into a few hours and everything must be precisely accurate.
>
> When I first started in the airplane business in Bristol, England, not very much was known about aerodynamics. But when something did go wrong, there was always a man up in the cockpit who could at least try to make things operate correctly. Such is not the case in the missile business. There's no man up in the cockpit, and reliability becomes a vital necessity. You can't have nuts and bolts that are not properly tightened, or electrical connections which are not properly soldered. Reliability is really important.
>
> Hundreds of tubes and gadgets and electronic contraptions of one kind or another are involved. This missile firing business is extremely complex in the testing phase, and the main purpose of the count down is to check to be sure everything is in place and operating properly. Loose wires or loose connections can cause misfirings.
>
> The last five minutes of a count down is much like walking the maternity ward when your first child is born. You wonder if the doctor is there, and you wonder what's going to happen, and the minutes seem like hours, and you wait, and wait, and wait. It seems like years.
>
> This is typical configuration of a modern missile or rocket: It is about 60 feet high, weighs from two to 20 tons, and probably has two or three engines. The second part has the brain. About 70 miles up, the brain tells the first part to blow itself off. It is traveling about

12,000 miles an hour at the time. Another engine is now started, and that continues up until it gets to about 300 miles. Then the end pieces and other excess baggage are blown off until only a comparatively small ball or projectile is left. This is what orbits at 500 or 2500 miles out, at about 18,000 miles an hour, and we expect it to stay up there for a period of from five months to 200 years. From these orbiting spheres or projectiles we accumulate scientific information. That is the simple way a missile fires: simple, yet very complicated. And that is why we go through a count down.

Some have said that the satellite proves there is no God. To me it proves the exact opposite.The only reason that we can orbit is because God has set up some rules and regulations and laws. The same force that keeps the stars going keeps the satellite going, when it is set in its right orbit. Only when it picks up a lot of dust, develops a drag and slows itself down does it come back to earth again.

This proves one thing - that God in the beginning created the heavens and the earth, and He did it by the word of His power."

I can certainly relate to Mr. Willey's reference to the importance of "reliability" in the missile or rocket business. As soon as we completed the Pershing program I was assigned to the quality assurance and reliability effort for a portion of the Apollo Program ground support equipment. My NASA managers anticipated my training needs and sent me to the proper schools and courses. Then on January 28, 1972, I was certified by the American Society of Quality Assurance as a Reliability Engineer. I devoted the rest of my career to this effort as we began the build-up of the Saturn launch facilities and the necessary ground support systems.

My duties included the placing of reliability requirements into the NASA/KSC contracts, the monitoring of the contractors who developed and provided the contract end items, and the final acceptance of the finished items. I visited many industrial plants across the country and met many fine engineers, technicians and managers who placed emphasis on the reliability and quality assurance requirements which resulted in the installation of the most reliable hardware ever assembled by any agency of our government.

Many of my prayers went into this effort as I gathered data which assured myself and my managers that we were receiving the kind of hardware that could be man-rated for our orbital programs and for our launch to the moon. Time proved that God answered my prayers because not a single count down scrub was attributed to bad ground support systems during the Apollo Program.

Another prominent scientist/engineer who had a great faith in God was Walter Francis Burke, former president of McDonnell Astronautics Company. When he was awarded the "Christus Vivit" award from Concordia Seminary (Lutheran, Missouri Synod) in May of 1966, the citation described him as "leader in the exploration of space, director from planning to launching of the Mercury and Gemini spacecraft projects, administrator, scientist and thinker; active and resourceful in service to his church, a gifted member of the body of Christ." The September 1967 issue of *DECISION* magazine published the following article by Mr. Burke entitled, "WAY OUT THERE:"

> "The scene is duplicated before every launch. The missile control center buzzes with technical talk, last-minute instructions, noisy chatter and pervasive American humor. Then at 30 seconds before lift-off the personnel seemingly begin to lose contact with the life around them. At T-zero some men about the room assume their personal attitudes of prayer. Some bow their heads, others clasp their hands tightly together, a few stand with eyes closed and lips moving inaudibly.
>
> I am among those in prayer. Things are out of our

hands and I pray for the safety of the astronauts. I ask that God's will might be done in the mission.Then when the rocket soars gracefully into space, and later as we begin contact with the spacecraft and talk with the spacemen or "walkers," my thoughts still are dwelling, in a sense, on the Creator. For the farther we penetrate in space, the more I am confronted with the wisdom and the majesty and the omnipotence of God.

I am confident that we will one day land on the moon and travel throughout our solar system - and perhaps even beyond. After all, we are merely visiting what God has created. What difference does it make whether we explore new frontiers by horseback or soar beyond distant horizons by rocket ship?

We feel something like this: "God knew that man would look into space - that's why it is so beautiful out there." He knew, too, that if we look, the next thing is to go. But it is impossible for anyone but God to know how far He will permit us to venture. I don't think we are tempting God one iota by going into space; otherwise He would have stopped us long ago." (Reprinted with Mrs. Burke's permission) (18)

Mr. Burke's observations in the T-30 second time frame have also been my observations during many of the Redstone, Jupiter, Pershing, Apollo and Shuttle launches. It was especially true when manned launches were involved. All of us were concerned that precious life was on board the vehicles. We lost all control over their lives as soon as the vehicle lifted from the launch pad, giving one such a feeling of helplessness that we just turned to God for His help and blessing as the flight progressed upward toward the goal of earth orbit, or to the moon, as was the case in the Apollo program. Many of the launch team members asked their churches, prayer groups and family members to say special prayers for the launches, and we launch team members felt those prayers as the excitement of T-zero approached.

I will never forget the Palm Sunday on April 12,

1981 when we launched Astronauts John Young and Robert Crippen into earth orbit in our first Space Shuttle, STS-1. It had been a beautiful countdown and a beautiful morning to launch, although I had to judge this from my console in Room 1P9 on the first floor of the Launch Control Center.

We could not leave our stations during the last thirty minutes, but we could all take a few moments to say a special prayer for the mission, and many of us did. I had a special feeling that this would be a good launch, because I knew that it was undergirded with the prayers of our nation.

It was a good launch and a good mission for which many of us thanked God. Included in giving thanks was our Launch Director, <u>George Page</u>. A few minutes after lift-off, while most of us were still at our consoles, George broke in on the intercom and loudspeaker system at the Launch Control Center and congratulated all of us for a successful launch and while he was speaking he said: "You know, we had a little help from the Lord." He then gave the microphone to Dick Smith, our Center Director, who had this to say as he was congratulating us for the successful launch: "We thank the Good Lord for a good Palm Sunday and a good successful launch." These words of thanks to God really inspired us as we secured our consoles and places of work - giving us all a good feeling that God truly answers prayers in man's endeavor to visit space and beyond. George Page is a distinguished engineer who came to work at Kennedy Space Center in 1963. He held various important Center positions including Launch Director for the first three Space Shuttle flights. While director of Space Shuttle Test Operations at the Vandenberg Launch Site he received the 1986 Elmer A. Sperry Award for his personal contribution to the world's first reusable Space Transportation System. The Award recognized "A distinguished engineering achievement...advancing the art of transportation." In my last chat with George, I reminded him of what he said during the congratulations session, and he shared with me his habit of stopping along the road and having a moment of prayer on his way to the Center for the critical tests and launches.

Richard (Dick) Smith began his tenure as Kennedy Space Center's director in September 1979 and held the post until July 31, 1986, when he retired from NASA to become the president and chief executive officer of General Space Corporation in Pittsburgh. He guided the space center during all our 25 shuttle launches until the loss of Challenger on January 28, 1986.

I have known this great leader since 1953 when we met at Cape Canaveral on the first launch of the Redstone rocket. He was a young Electrical Engineer and I was a young Electronic Engineer engaged in preparing the nation's first ballistic missile for launch. Most of his work was at the Marshall Space Flight Center in Huntsville, Alabama where he served as Deputy Director before going to NASA Headquarters to lead the Skylab reentry program for NASA. Dick has a great faith in God as shown in the statement above. The Smiths attended the Grace United Methodist Church of Merritt Island, Florida.

RICHARD SMITH (left) AND GEORGE PAGE, 1981
Congratulating each other on successful
launch of STS-1 (NASA PHOTO)

More can be said about prayer for our space launches as reported in the July 24, 1997 issue of Florida TODAY newspaper. The article was about the prayer support given to the Air Force launch teams at Cape Canaveral. The Florida TODAY representative asked me to go directly to Chaplain Phillip Nguyen for permission to use the following prayer which was placed on each of the consoles at the Cape Canaveral Range Operations Control Center during the recent countdown for a GPS IIR-2 atop a Delta launch vehicle:

> "Almighty and Gracious God, we praise you and give you thanks for the blessings of creation, knowledge and wisdom. As we look to the beach...and beyond (Where No IIR Has Gone Before), we seek your divine guidance and invinceable guardianship, for the safety and success of this vital mission. You have granted us the materials and the ability to accomplish many great tasks. Now we ask that you would grant us fair weather...calm winds and clear skies. O Lord, just as we ask for your blessing upon this evening's Delta launch and pray for the safety of all involved, we come here this evening in humble meditation. May it be your continuing presence and infinite wisdom which directs us in the decisions we make and the actions we take. May the success of this launch and the use of GPS IIR-2 bring honor and glory to your Holy Name. Amen."

Chaplain Nguyen, the Roman Catholic chaplain graciously gave me permission to use this prayer in my book. It is my hope that the reader will be inspired to continue praying for each launch for both the Air Force and NASA.

One can see that all of society has gained numbers of us who believe in prayer. There is more interest in Guardian Angels and all Angel lore throughout American society. The media has taken more interest as we can see on television and at the book stores. Are we finally believing that "He has the whole world in his hands?"

The first director of the John F. Kennedy Space Center was Dr. Kurt Debus. He was known as the architect of the the nation's spaceport. Dr. Debus was a member of the famous Peenemunde Research Center that designed the V-2 rockets with which Germany bombarded British and Belgian cities during World War II. He was one of the 130 German scientists, engineers, technicians and administrators who brought their expertise to work for the U.S. Army at the end of the war. Dr. Debus was instrumental in the selection of Cape Canaveral as the nation's major space launch site. He personally directed the launch of the first American satellite, the first American satellite in solar orbit, and the first flights of primates in Jupiter and Mercury spacecraft.

Dr. Debus also oversaw the lofting of the first American into space; the flight of Alan Shepard, Jr. on May 5, 1961; launch of the first humans to the moon; and launch of Pioneer 10, which became the first probe to leave the solar system.

Dr. Debus was born in 1908 in Frankfurt, Germany. He earned his engineering degrees from Darmstadt University. He came to the United States in 1945 with Dr. von Braun's team and right away was chosen to lead the launch teams at White Sands and later at Cape Canaveral. I first met him in 1952 and was one of the first ten men chosen to accompany him to the Cape to launch the first Redstone rocket. The Redstone was derived from the research conducted at White Sands and closely resembled the V-2 in many ways. His first effort was to build a launch team. Through the efforts of this team, an efficient and effective launch team was organized and trained to perform various launchings which put our country into the lead in space exploration for many years.

In order to boost mankind into space, special launch vehicles and spacecraft had to be developed, tested and flown. This talented team of engineers and technicans pioneered the way for accomplishing the successful testing and launching of the Redstone and Jupiter missiles, the Army Pershing missile, the Saturn rocket series, and the Space Shuttle.

Dr. Debus was at the forefront in all facility planning and design effort. I remember how this shy genius would sit with me and show me how he wanted a particular report finished, including content and format. All of the launch team appreciated the time and attention he paid to our work. He expected the best of all of us because he always put his best into his work.

DR. KURT DEBUS AND PRESIDENT JOHN F. KENNEDY, Nov. 16, 1963 (NASA Photo Taken During President Kennedy's Visit To The Cape)

DR. KURT DEBUS Inspects the Silver Beaver Award presented by the Boy Scouts of America to the Author in 1963. {NASA Photo}

Dr. Debus was a devoted husband and father to his family and showed interest in each of our families as we continued to build a close-knit team, and as we settled into our community and church activities. I felt that he always respected me and supported me as I led a Boy Scout Troop at my church. When I received the Silver Beaver Award from the Boy Scouts of America in 1963, he called me into his office to see the award and to congratulate me, as shown in the preceding picture. He trusted me to host most of the youth groups who came to the Cape on tours, which I was proud to do when my duties allowed.

Dr. Debus did not attend a church, but those of us who knew him and worked closely with him had a good feeling about his relationship to his Maker. This feeling was corroborated at his funeral in October, 1983 by his daughter's Baptist friend, Rev. Dr. Paul Allen. The following is taken from my personal recording of the funeral. Dr. Allen said:

> "I remember spending a long evening several years ago in Dr. Debus' home in Cocoa Beach, and we had the opportunity to discuss a number of things. Dr. Debus was not a member of any organized religion, nor as far as I know did he ever subscribe to any peculiar or particular tenets. As I talked to him that evening I remembered I was talking to a scientist of high reason. I came to feel that I was speaking with a man who somehow and in his own way, in his own manner, had come to reckon with his creator, the God of Heaven and Earth. I can understand that. If an astronaut is circling the globe and looking down upon the Earth, I can well imagine that he might shake his head and say, "I wonder if there is a God?", because as he flies around, he is flying over corruption, he flies over war; he flies over suffering, he flies over selfishness; he flies over every mess that man has made for himself in this beautiful world that God gave us. When he looks down that is what he sees. But if he stands on the Earth and looks up to the sky and sees the stars in their places and knows that there has to be a Creator, it is almost impossible for a man not to come to grips with that subject and somehow come to grips with his Maker. I believe that Dr. Debus has done that, and it is in that

spirit that this service is conducted today."

Mr. Grady F. Williams, who was my first supervisor when I came to Huntsville, Alabama, recently told me that Dr. Debus certainly believed in God. He remembered several instances in which this view was verified. Mr. Williams was a devoted Deacon in his own church and served as a role model for a lot of us space workers as the years went by. My wife and I were invited to his ordination in the fifties here in Florida. It was a most moving experience and was important in Grady's life as he continued to lead us in the various positions of responsibility at the Cape and later at Kennedy Space Center (KSC).

Mr. Williams served as KSC's Director of Design Engineering from 1970 to 1972 when he requested a change in assignment to concentrate on development of a launch processing system for the Space Shuttle. He also served as Deputy Director of Design Engineering from 1967 to 1970. After serving in the Air Force during World War II, he attended Auburn University where he received his BS degree in electrical engineering in 1949. He served as an instructor at Auburn, then joined General Electric as a design engineer on the Hermes II missile at White Sands, New Mexico and participated in the last launch of a V-2 rocket.

White Sands, New Mexico is where Mr. Williams joined Dr. von Braun's team and continued with them when they were transferred to the Redstone Arsenal at Huntsville, Alabama. He served as an electronic design engineer in the technical and engineering division of the Guided Missile Development Group. Later he was the chief of our Measuring Section of the Army's Missile Firing Laboratory at Cape Canaveral, Florida. After the Army transferred us to the National Aeronautics and Space Administration, Grady served as Chief of Electronic Engineering and Instrumentation Systems Division of the Launch Vehicle Operations Directorate. He then became the Deputy Chief of the Electronic, Measuring and Tracking Office of the NASA Launch Operations Center. Like many space pioneers, Mr. Williams did a lot of praying for the launches, especially for the manned ones. He felt that God did intend for man to enter the realm of space and to have dominion there. In conversations with Grady, I felt closer to God myself. He recently told me where he early learned to place his priorities: his first duty was to God, then to his family and then to his job. His life before his co-workers and friends certainly bore that out.

Grady F. Williams, Former Director of Kennedy Space Center Design Engineering, 1970-1972 (NASA Photo)

Another important leader who gives insight into his belief in God, and in his belief that mankind should be going into space is Dr. Robert A. Frosch. Dr. Frosch was nominated by President Carter as our NASA Administrator on May 23, 1977. He earned his Doctorate in theoretical physics from Columbia University in 1947 and began his career in marine and related geophysical sciences as research scientist, as director of research laboratories and various research projects, and as an executive of science-related activities in the Department of Defense and the United Nations. He was responsible for such projects as the underwater acoustic exploration of the Norwegian Sea; United Nations environmental programs in Nairobi, Kenya; and nuclear test detection and advanced research projects for the Department of Defense.

From Late 1966 to early 1973 he held the critical post of Assistant Secretary of the Navy for Research and Development, initiating important improvements in Navy research and development which earned him the Navy Distinguished Public Service Award in 1969. He holds many other awards, and he is the author of a number of professional papers.

I want to share what he told NASA employees in the December 1977 issue of *NASA Activities*, a periodical for all agency employees:

> Dear Colleagues:
>
> The heavens declare the glory of God; and the firmament showeth His handiwork. Day unto day uttereth speech, and night unto night showeth knowledge. (Psalms 19: 1,2)
>
> Throughout history, and before the history we know, the order and the glory of the planetary system and of the stars have been the sign of order in the universe and the inspiration of religion and of strivings for better order in human life.
>
> As we look outward on the universe, sending

DR. ROBERT A. FROSCH, NASA ADMINISTRATOR,
Appointed by President Carter on May 23, 1977
(NASA Photo)

instrument extensions of ourselves beyond where we can yet go, let us also consider how we may create a better order within and among us, so that the knowledge and power we gain may be used for the best advantage of us all.

We in NASA share the special privilege of working toward understanding the universe and learning how to use that understanding.

The year 1977 has been one of accomplishment for our agency, marked, among other things, by the continued success of the Viking Mars explorations, the completion of the Space Shuttle approach and landing tests, the initiation of the Voyager odysseys, the beginning of HEAO and ISEE operations, numerous significant technological developments for aeronautical and space applications, and the continuous flow of that technology into other sectors of society to the general benefit of the human lot.

As NASA people, as government employees, as Americans, we have every reason to celebrate a year filled with progress for our country, a progress to which we have personally contributed. My own satisfaction in this season is to be the leader of the legendary NASA team with its immense potential for the advancement of science and the improvement of human life on our planet. (19)

By the time Dr. Frosch became the NASA Administrator, I had already retired from NASA and was working with a support contractor in my second exciting career at Kennedy Space Center. I had a great respect for him and followed his leadership with excitement as we prepared the space shuttle ground support launch equipment and participated in the other NASA vehicle launches. He certainly believed in God and in His leading of mankind into space and in working toward understanding the universe and learning how to use that understanding.

Another great engineer who believed that God would let mankind enter into space was Dr. Christopher C. Kraft. Chris Kraft was always in the forefront of space activities from the early Mercury program. The May 14, 1972 issue of the *FloridaTODAY* newspaper gives us some insight into his faith in God:

> Before John Glenn became the first American to orbit the earth in 1962 he said he wasn't scared because he knew "so much about it."
>
> Ten years later the man who then knew more than Glenn still prays to God when astronauts return from the moon and slam into earth's atmosphere.
>
> Dr. Christopher C. Kraft, 48, developed the equipment, trained the men and personally directed every Mercury flight and half the 10 Gemini flights. He also designed Mission Control in Houston which controls Apollo moon missions.
>
> Now Kraft is director of the Manned Spacecraft Center in Houston following his appointment in January.
>
> On the splashdown day of the Apollo 16 mission he was asked how his day began.
>
> "The first thing when I woke up I said a prayer to God that the control systems and recovery equipment would function, and that we would see three good parachutes.
>
> You have your heart in your mouth, waiting for a successful recovery to happen - thanking God when it does," Kraft said.
>
> His glass-walled office on the top floor of the MSC headquarters building affords a spectacular view of the center and the skyscrapers of Houston 25 miles in the distance.
>
> Pieces of heat shields from several Mercury missions encased in plastic occupy one corner of the office. On his massive desk the small status board that had to flash green before a Mercury launch could take place now holds two pens.

DR. CHRISTOPHER C. KRAFT, Former Director
Johnson Space Center (NASA Photo)

"I'm amazed at the ability of the entire crew to carry it out. I'm even concerned on an airplane in rough air, hoping it will remain structurally sound," Kraft said.

A visitor expressed surprise at Kraft's concern over recovery of Apollo 16.

"It is the normal response of a man who knows the complexities of everything. It is the nature of an engineer who understands what has to function to make it work. I once said if you took all the probabilities of failure into account you'd never go to the pad," Kraft said.

"There's going to be a risk to the safety of astronauts for a long time, even with the space shuttle. It will be a long time before you have the kind of reliability you have with the commercial airlines," he said. (20)

Another important aerospace engineer was my good friend, Mr. Alfred D. O'Hara. Al got his start in the space program at Redstone Arsenal as an Army Lieutenant with our Von Braun launch team in the early fifties after I had joined. He studied for his degree in electrical engineering at the University of Miami. After four years in the Army he joined the Missile Firing Laboratory here at the Cape. I watched the progress of this young engineer with admiration and respect.

Al eventually became technical assistant to Dr. Hans Gruene, another of the rocket pioneers who came to this country with Dr. Von Braun and was assigned to lead the ground operations unit under Dr. Debus. Al joined us in time to begin work on the Saturn 1B and was operations manager by the time the first one was launched into earth orbit. Al, a native of West Palm Beach, readily attributed his steady rise in the space program to his Baptist faith. He told the media that he was a "born again" Christian and said that "without the Lord's help, I'd be useless."

Al had served his church as a Deacon, Sunday School

ALFRED D. O'HARA (Left), Former shuttle launch director (NASA Photo)

Teacher, and Chairman of various committees for over two decades.

Even as Saturn 1B pushed a 20 ton payload into earth orbit, Al became test and operations manager for the 363 foot Saturn 5, the mightiest of all American rockets. Apollo reached its historic climax July 20, 1969, when Neil Armstrong and Edwin Aldrin stepped onto the lunar surface.

At the end of the Apollo program in December 1972, Al took time out to earn a master's degree in education and administration at Stetson University.

In June, 1982, Al replaced George Page and became launch director for the Shuttle Program and led our team as we launched the Shuttle Columbia twice and Challenger on its debut in April, 1983.

When asked by the media if he found conflicts between work and faith, he replied, "None, my priorities are faith in God, family, church, and work." But he did admit, that like most of us, he didn't always keep them in proper order. "Work rates too high at times," he said.

It was my privilege to assist Al on a major procurement for the Shuttle Checkout, Control and Monitoring System. This critical launch preparation and launch control system proved itself well during the Apollo years. One begins to know a person for who he is as one works day-to-day and shares thoughts and ideas. I know that Al did a lot of praying for each test and for each launch, and put his trust not only in the 7,000 civil service and contractor personnel who were working under his direction, but in the God we all trust at T-0.

Another instance of asking God's blessing on a satellite launch was when the engineers of Dr. Robert Gray's Vanguard launch team submitted a drawing change (DCR-1000) which allowed for a small Saint Christopher's medal to be safety-wired and taped to the flight guidance package. The paper work was officially signed off and the reason given at the bottom of the Drawing Change Request was: "Addition of Divine Guidance." After the successful launch of Vanguard 1 on St. Patrick's Day, March 17, 1958, Dr. Gray commented that: "I don't know, maybe it worked." God surely blessed that effort, because the tiny satellite is in such a perfect Earth orbit that it may remain in space hundreds or even thousands of years.

I personally asked Dr. Gray about this instance, and he corroborated it as having actually happened among his team. He speculated that God did honor that request couched in the Design Change Request, DCR-1000.

Dr. Gray and I are members of the Florida Chapter of the NASA Alumni League and enjoy the fellowship of our retired Kennedy Space Center friends at monthly luncheons in Cocoa, Florida.

As I look around at my fellow space pioneers at these meetings, I feel a definite kinship with them. Many of them were key managers of space flight activities and contributed to the U.S. space race and accomplishments that benefit all of mankind. What an honor it has been to work among them and share such a great sense of accomplishment in expanding man's domain into God's universe. Their prayers counted in all progress made, and I feel sure that many of them still pray today as space travel continues toward our country's goals in space.

According to the Associated Press the Russian Orthodox Church has started a praying tradition with the Russian Cosmonaut launchings. A black-robed Priest did the first-ever prayer service at the cosmonauts' hotel before their flight to the MIR Space Station which was to lift off on Tuesday, August 5, 1997.

This was to be a very dangerous mission for the cosmonauts, Anatoly Solovyov and Pavel Vinogradov, who had been in training for the extensive repairs planned at the space station, which was damaged by a cargo supply ship on June 25, 1997.

One can see the hand of God at work in Russia as this one act of open praise and dependence on the one true God for His blessing on a state venture such as the Russian Space Program was performed and announced to the whole world. Praise God and praise His many followers in Russia!

Chapter 5. ASTRONAUTS BELIEVE

During 1958, scientists and engineers, both military and civilian, talked more openly with each other and with the Congress about the possibility of launching man into space. Much experimentation with lesser species had been conducted with much confidence being gained that man could withstand the rigors of the space environment. They talked more openly about radiation dosages, meteoroid penetration, weightlessness and the other anomalies of space travel.

What made members of Congress and the American people responsive to such discussions were the ever-larger scientific satellites launched by the Soviet Union that began on October 4, 1957. In the midst of the nationalistic humiliation following the Sputniks, not only space rocketry, but also medical research with rockets, received an invaluable boost. Beginning in December 1959, the Navy School of Medicine dispatched monkeys, sea-urchin eggs and sperm, molds, tissues, and seeds on our Jupiter missile at the Cape. I was heavily involved in the preparation and launch of these vehicles and felt that we were indeed making history and making progress toward finding a place in space for mankind.

One setback in my confidence happened on December 13, 1958 when we launched Gordo, a small squirrel monkey, on Jupiter AM 13. In the preparation for this launch, two external measurements were canceled, and one of my measurement technicians was instructed to coil up the cables from these measurements and secure them inside the aft section between the stages. It so happened that the cables should have been fastened to the lower bulkhead of the upper section, but were secured with lacing cord to the lower stage. So when separation came, the separated upper stage moved forward until the length of secured cable caused it to stop and go out of control. This caused the internal stabilized guidance platform to tumble and we lost the mission. The person in charge of this preparation and I took the responsibility for the loss of this mission and I still hurt today knowing that I was partially responsible for that failure.

Such human error was inexcusable but ever present in launch mission preparation as the space program progressed. Much data of this type was accumulated before the managers agreed that a good quality assurance program would be needed before we could propose a manned flight. I was instrumental in helping set up the original reliability and quality assurance program at the Cape and continued in that role for over twenty-five years.

I wrote a "History of Quality Assurance at Kennedy Space Center" in 1974 which covered the period of the early fifties to the end of the Apollo program. The human error factor in my research showed that out of 10,000 case documents, 7400 involved human error as a contributing factor. The government decided to proceed with a man-in-space program, while being determined to markedly reduce such human error.

The first effort was called "Man in Space Soonest" and was the brain child of the Air Force and the National Advisory Committee for Aeronautics (NACA). The Army tried to justify a similar project called "Man Very High" but failed. The Army then changed the name of their proposed project to "Project Adam." Much planning and research was going on within the Air Force and the Army circles while final plans were being negotiated for a manned satellite. The National Aeronautics and Space Act of 1958 finally decided that NASA would be given the job of directing the manned satellite program, and the name of the program was changed to "Project Mercury".

In parallel with hardware development came the job of selecting astronauts to train for the space ventures that were being planned as the program began to take shape. NASA organized the Space Task Group at Langley to lead in the astronaut selection and training program. After much wrestling with the selection criteria, the following seven-item formula was followed:

1. Age - less than 40.
2. Height - less than 5 feet, 11 inches.
3. Excellent physical condition.
4. Bachelor's degree or equivalent.
5. Graduate of test pilot school.
6. 1500 hours total flying time.
7. Qualified jet pilot.

When these criteria were given to the Pentagon, service record checks revealed that more than 100 men on active duty appeared to be qualified.

The military services were pleased to cooperate in further screening. NASA was relieved not to have to issue an open invitation to the public sector at this time and the selection process began.

An air of national excitement followed the selection process, and the whole nation was awaiting the names of the first astronauts. Seven were chosen, and the whole nation knew their names by heart within a short time because of the intense media concentration on them. They were Malcolm Scott Carpenter, Leroy Gordon Cooper, John Herschel Glenn, Jr., Virgil Ivan Grissom, Walter Marty Schirra, Alan Bartlett Shepard, and Donald Kent Slayton.

An intensive training process began immediately. All seven astronauts had come into the Mercury Program with aviation flight training behind them. Now they must go through training in astronautics, space biology, conditioning for space flight (including centrifuge and closed loop analog static simulator), participation in the Mercury spacecraft development plans, operation of the Mercury vehicle, and familiarization with ground operations, both for the Redstone and the Atlas launch vehicles. Every contingency was thought of and researched and studied by space medicine specialists and psychiatrists in anticipation of the stress that was in store for human beings subject to flight into such a hostile environment as outer space. Special emphasis was placed on the effects of isolation. It was about this time that reports came through of our captured soldiers in North Korean prison cells being "brainwashed" so easily, with isolation as part of the process.

In 1966, NASA gave all Mercury Program participants a copy of the book, "This New Ocean" from which I will quote starting on page 232, that concerns the isolation tests performed:

> "During the fifties academic and medical studies in sensory deprivation made an important, if indirect, contribution to the building of the spacecraft and the training of the astronauts. Made notorious by the experience of American prisoners of war who had been isolated and "brainwashed" in North Korean prison cells, the effects of isolation were attacked on many fronts. At McGill University in Canada, at the University of Rochester, New York, and at the National Institute of Mental Health in Bethesda, Maryland, the sensory deprivation experiments reduced all physical stimuli to near zero. Suspending people in water of body temperature in blacked-out, soundproofed rooms at Bethesda revealed that normal men, regardless of their motivations, could hardly stay both conscious and sane if deprived of all sensory stimuli beyond three hours. Physicians and psychiatrists were warning in 1956 and 1957 that if "one is alone enough and at levels of human and physical stimulation low enough, *the human mind turns and projects outward its own contents and processes........"* (Italics mine) (21)

It's at this point where man's inner makeup comes into play. The point being that man needs help from within to project outward the correct contents and processes needed for isolated flight into the unknown. What better source for such help than from the Word of God. Matthew 15:16 gives us knowledge that must be heeded:

"But those things which proceed out of the mouth come forth from the heart; and they defile the man."

My feelings are that all the training and conditioning given to the astronauts was of itself not enough to thoroughly convince me that they would be able to perform rationally for the expected time period of isolated flight without a deep faith in God. Faith in God must be folded into the needed social science for long space flight.

The July/August, 1995 issue of "AD ASTRA," the magazine of the National Space Society, page 55, projects some thinking along this line:

> "Today's space program tends to steer away from social science and its effect on space missions, both near-term and long-term. The crews of the Russian MIR space station and its predecessors fought with each other or disregarded ground controller's instructions on several occasions when the stress of living and working in a confined environment in space for an extended time got to them. Small groups of researchers living in places like Antarctica or sailors aboard submarines have exhibited aberrant or antisocial behavior after living in isolation....Social science is very important to getting work done on a mission."(22)

The November 1988 issue of National Geographic gave some thoughts on what a crew to Mars might face: "A Mars crew faces one final hurdle: the psychological double whammy of prolonged isolation and confinement. Isolation from all they have known before: no family, no trees, no valleys, no waterfalls. Out of the windows, month after month, the same black velvet sky dotted with unblinking stars; only the almost immeasurable change in sun, earth, and Mars as the mission unfolds.

And confinement: no way to escape from the cloistered daily life of the spacecraft. No way to avoid looking at the same walls, no way to escape from the same repetitive tasks, or from other crew members. The interaction between individuals may be the most critical factor of all, as months of boredom creates frustration and hostility, followed ever by incapacitation or violence.

Perhaps the closest analog on Earth is wintering over at an Antarctic research station. For the scientists and the supporting staffs, the problem often results in "cocooning," - a form of withdrawal. Cases of violence have been reported. Under such conditions, organization and compatibility are of paramount importance. Similarly, reports from long-duration Soviet space flights tell of tiffs between crew members, with participants on 2-man crews pouting at either end of their space station.

Astronaut Norm Thagard exhibited a lot of isolation stress on his recent stay on the Russian MIR space station and did not hide his complaints from NASA. The Russian cosmonauts described Norm's complaints:

a. The lack of news and family contact during his 115-day mission.

b. His distaste for Russian space food.

c. He didn't have enough contact with English-speaking people.

Al Martin, an American who was in the middle of his third winter-over at McMurdo Sound, Antarctica, says that NASA should heed Thagard's request that the psychological as well as physiological needs of astronauts be dealt with during extended periods of isolation. He feels that the psychological aspect is probably the more important. (comments taken from The July 23, 1995 issue of Florida TODAY newspaper).

Certainly, religion plays a great and favorable part in the psychological make-up of people, and we must not ignore this element in personality building.

Religion seemed to play a great part in the day-to-day well-being of those in isolation in both Korea and Vietnam.

Thoughts of God seemed to give each Prisoner of War (POW) a sense of direction as they later reported in interviews or in their writings. They felt relief from uncertainty, frustration and inner conflict that brought confusion to many of their fellow prisoners. Those who declared a faith in God really had "right stuff" before author Tom Wolfe popularized the term.

Author Tom Wolfe reported on the religious faith of the original astronauts as a component in their "right stuff." I could agree with Tom on this because my research may have preceded his studies.

Female astronaut, Shannon W. Lucid, had a degree of “right stuff” also. She is the daughter of the Baptist preacher Joseph Oscar Wells and Myrtle Wells, a missionary nurse, who were stationed in war-torn China where Shannon was born in 1943. The family was interned in a Japanese prison camp for a year, then released in exchange for Japanese POW’s. Her family returned to China but had to leave when the communists took over in 1949.

The October 7, 1996 issue of NEWSWEEK reported that Shannon had a love for science with a focus on the heavens. She earned her pilot’s license and flew her father to revival meetings in an old Piper Clipper she bought. Her dad recalled her saying, “The Baptists wouldn’t let women preach, so I had to become an astronaut to get closer to God than my father.” So we can say with assurance that Shannon depended on God to care for her throughout her space adventures and helped her cope with the crew members locked in small spaces for 188 days aboard the MIR Space Station.

The December 30, 1994 issue of The United Methodist Review reported an interview with church member Susan Helms who was named an astronaut in 1990. Ms. Helms was on the Space Shuttle Discovery mission September 9-20, 1994. A portion of the interview follows: “.......The Discovery mission was the second space flight for Ms. Helms, named an astronaut in 1990. She said her faith did not enter into ‘the pros and cons’ of whether or not to go into the space program, but you don’t step onto a controlled explosion like that without having some kind of faith that there’s a reason for everything, a master plan.'

She continued, 'If you're afraid of death, you certainly wouldn't do something like this, but if you have a spiritual life and do not fear death, you're not afraid to attempt it.'

Ms. Helms said she practices her faith on a daily basis like anyone else. 'Nothing about being an astronaut makes us special except that we get the stupendous opportunity to look back on the earth pretty much the way God made it and can appreciate it in a way most people don't get a chance to do,' she said."

Astronaut David Leestma, featured in the August 1977 Home Life Magazine's article by Art Toalston, LAMP UNTO MY FEET, expressed his faith when he said: "In my line of work you can start depending a lot on yourself. But if you continue depending on yourself, you're going to go astray. You forget that God's in control.....Proverbs 3:5-6 always bring me back to what's really important: everything works out a lot better when you trust in God, acknowledge that He's in charge and follow His direction............"

Future social science studies should seriously consider God, the Creator of the universe, as plans are made to groom mankind for the dangerous ventures into His domain, calling for longer and longer periods of isolation. Certainly the studies should do this, if groups of people are to function in love and harmony.

With the announcement of the seven astronauts, I began to seek out and collect statements they made to the media, or in their books, about their faith. Over the years I have been able to fill scrapbooks with such statements of faith, not only of the original seven, but for the Gemini, Apollo and Shuttle astronauts as well. I now want to share some of their statements with the reader:

ALAN B. SHEPARD, Relaxing with the flight team prior to his historic flight aboard Mercury Redstone, MR-3, May 1961. (NASA Photo)

"ALAN B. SHEPARD, JR. Was born in East Derry, New Hampshire, November 18, 1923. He received a bachelor of science degree from the U.S. Naval Academy in 1944.

Following destroyer service in the Pacific in World War II, he entered flight training and received his wings in 1947. Shepard's assignments in the Navy include U.S. Navy Test Pilot School at Patuxent River, Maryland, and flight test work there. He participated in experiments in development of the Navy's inflight refueling system; carrier trials for the F2H3, Navy trials of the first angled carrier deck; testing of the F3H, F8U, F4D, F11F; and project test pilot on the F5D. He graduated from the Naval War College, Newport, Rhode Island in 1958 and was assigned to the staff of the Commander-in-Chief, Atlantic Fleet, as air readiness officer.

Shepard was one of the seven astronauts selected by NASA in April 1959 and is now Chief, Astronaut Office. On May 5, 1961, he was this country's first man in space as pilot of the Mercury-Redstone 3, a suborbital flight. The flight attained an altitude of approximately 116 statute miles and the spacecraft traveled about 302 statute miles down the Eastern Test Range.

Shepard was awarded the NASA Distinguished Service Medal in 1961, and the Langley Medal in 1964." (23)

The April 20, 1986 Florida TODAY newspaper conducted an interview with Alan to which I refer for Alan's statement on his faith in God:

> **Question:** I know that you and Louise are Christian Scientists. Did you have any religious thoughts of, or say any prayers before, during or after either of your space flights?
>
> **Answer:** "No, My belief in God has always been a day-to-day thing -- that He was right here on Earth and being a quarter of a million miles away didn't bring me closer. Others have different feelings and

approaches. Mine was always associated with a God that helps me on a daily basis. You know, up is down if you're on the other side of the Earth. I've always felt that my religious help came here and wasn't enhanced or degraded by being in space." (24)

The initial manned launch from Cape Canaveral did not get off without God's blessing being officially invoked. Dr. Kurt Debus and his team had arranged for a local pastor to give a brief prayer at blockhouse 5/6, and a Navy Chaplain aboard the carrier Lake Champlain gave the following prayer that was reported in "WE SEVEN," published by Simon and Schuster, New York, N.Y. 1962, page 179:

> "Gracious God," the Methodist chaplain said, "as a precious human life is propelled into the heavens, we become apprehensive, tense and fearful from anticipation of imminent danger. We are thankful there are men who in their dedication are willing to prepare our way and open the door into space by sacrificing their very lives if necessary. May success crown our efforts to explore and develop paths to achieve not only an expanding universe, but a safe and more peaceful one in which we may live with ourselves and with Thee, Amen." (25)

In their retirement years, Alan and the other original astronauts are active in their own organization called "The Mercury Seven Foundation," which sponsors the Astronaut Hall of Fame. The Hall of Fame is the home of the U.S. Space Camp in Florida where children are encouraged to study math, science, and technology, in a fun, hands-on learning environment. This is positive evidence of the original astronauts' faith in action.

The U.S. Space Camp Florida continues to grow and has become the chief endeavor of Mercury Seven Foundation and the U.S. Space Camp Foundation. I visited the first Space Camp at Titusville, Florida in March 1987 when I was asked to present my paper, THE STORY OF SPACE FLIGHT AND KENNEDY SPACE CENTER. I found that initial team of counselors to be very interested in my talk and they grouped around me for several minutes afterward to ask more specific questions in their preparation for the first Camp session at Titusville.

I was present at the ground-breaking for the Titusville Space Camp. Thousands of young people from all over the world have benefited from their visit to this great undertaking, instilling in them a stronger interest in science, math and space travel in general. May SPACE CAMP continue with just that goal, both at Titusville, Florida, and at Huntsville, Alabama, as the years go by, because we know that the future of space science and space travel will be part of our young peoples' heritage.

VIRGIL I. (GUS) GRISSOM,. Second American In Space, Mercury-Redstone 4, July 21, 1961. (NASA Photo)

"VIRGIL I. (GUS) GRISSOM was the "Old Man" of the Apollo 1 crew. Command Pilot Grissom was one of the seven original astronauts who started the nation's space program in April of 1959. He was a space veteran of the Mercury-Redstone 4 (Liberty Bell 7) suborbital mission on July 21, 1961 and served as command pilot of the first Gemini flight on March 23, 1965, with John Young as pilot.

Grissom performed the first orbital maneuver ever conducted by man in a spacecraft during flight. He was named command pilot of the Apollo 1 crew on March 21, 1966.

Grissom was a much-decorated hero of the Korean air war, where he flew 100 combat missions in F-86s with the 334th Fighter-Interceptor Squadron. He wore the Distinguished Flying Cross and the Air Medal with cluster for his Korean service; two NASA Distinguished Service Medals. the NASA Exceptional Service Medal, Astronaut Wings and the Firefly Club Award." (26)

Just after the astronaut selection was ended, the NASA Administrator, T. Keith Glennan, held a press conference for the astronauts so that everyone in America could get behind them and the space program. It was at this conference that we first learned about the religious faith of the astronauts. When it came Gus Grissom's turn to make a statement, he said that he considered himself religious. He said that he was not very active in his church, but did consider himself a good Christian. His mother always took him to church and this eventually gave him a good foundation in Christian faith.

Our Team Presenting Gus Grissom the "Good Guy" Award in 1961

We on the Mercury-Redstone team did not get to see Gus very often since he was in intensive training at Langley or in Hangar "S" at the Cape. I do remember the occasion when the Space Task Group and the Debus team gave Gus a "Good Guy" award. The award was a sizable desk lamp set (shown above) which we thought he could remember us by. He was indeed a "good guy" and easy to work with. His attitude and goodwill gestures endeared him to all of us on the launch team. This picture puts Gus in the middle of the front row near his desk lamp, Alan Shepard second to his right; myself at the right end of the second row (checkered coat); and John Glenn, second from left on the back row. Other men pictured continued to advance rapidly in the NASA ranks to positions of greater responsibility within the space program.

On January 27, 1967, astronauts Grissom, White, and Chaffee gave their lives for the important goal of landing Americans on the Moon. It is hard to describe the hurt we felt when word was released of this tragic accident. Maybe if NASA and North American Aviation (NAA) had listened more carefully to one of the NAA Quality Control Specialists about his concerns for the safety of the astronauts, and had given more credence for the discrepancies he reported, this catastrophe would never have happened with astronauts on board.

I met with the NAA Quality Assurance Specialist, Mr. Tom Baron, unofficially on December 19, 1966 at the Holiday Hospital in Orlando, Florida, while my son was there for surgery. This was before the accident on January 27, 1967, and I listened to Baron's concerns and told supervisiors of my feelings. I was hoping that close attention would be given to his allegations of poor workmanship and loose follow-up on his Discrepancy Reports (DRs). I prayed that a 100 percent oxygen test would be performed without the crew. I sometimes regret that more of us did not put our jobs on the line for better quality and safety practices concerning spacecraft 204. Although Mr. Baron could not prove all his "Good Faith" findings to be correct, his forebodings proved correct and three great American astronauts lost their lives in the Apollo 204 fire.

The whole nation and government leaders suffered along with the families as the Apollo Program was reviewed many months with many improvements made to the Apollo spacecraft before resuming operations at the Cape. I am sure that this was one of those times when NASA management wishes it had been more interested in safety and quality concerns rather than in pushing to meet schedules.

Our Director, Dr. Kurt Debus, as candidly as possible, with great feeling, gave this short statement to the Congressional Investigators:

"As director of the installation I share the reponsibility for this tragic accident and I have given it much thought. It is for me very difficult to find out why we did not think deeply enough or were not inventive enough to identify this as a very hazardous test.

I have searched in my past for safety criteria that we developed in the early days of guided missile work and I must say that there are some that are subject to intuitive thinking and forward assessment. Some are made by practical experience and involved not only astronauts but the hundreds of people on the pads.....We never knew that the conflagration would go that fast through the spacecraft so that no rescue would essentially help....This is essentially the cause of the tragedy."

JOHN H. GLENN, JR., First American to Make an Orbital Flight, February 20, 1962 (NASA Photo)

"**JOHN H. GLENN, JR.** was born July 18, 1921, in Cambridge, Ohio and attended Muskingum College at Concord, Ohio. Entering the Naval Aviation Cadet Program in 1942, he was commissioned in the Marine Corps in 1943.

In April 1959 he was selected as an astronaut for Project Mercury. As pilot of "Friendship 7" spacecraft, Glenn was the first American to make an orbital flight. Launched from Cape Kennedy, Florida, the flight reached a maximum altitude of 162 miles and made 3 orbits in 4 hours, 55 minutes, 23 seconds, before landing in the Atlantic Ocean near Grand Turk Island. Awarded the Distinguished Flying Cross five times, Glenn also holds the Air Medal with 18 clusters. He retired from the Astronaut Program in 1964, and became a consultant to NASA's Administrator in February 1965." (27)

At the first astronaut press conference in Washington with Dr. T. Keith Glennan, John was leading the group in witnessing of his faith. Those Christians present must have been excited when John made this statement:

> "I am a Presbyterian," he says, "a protestant Presbyterian, and I take my religion very seriously, as a matter of fact. I was brought up believing that you are placed on earth here more or less with a sort of a fifty-fifty proposition, and this is what I still believe. We are placed here with certain talents and capabilities. It is up to each of us to use those talents and capabilities as best we can. If you do that, I think there is a power greater than any of us that will place the opportunities in our way, and if we use our talents properly, we will be living the kind of life we should live."

After John's orbital flight in Freedom-7 on February 20, 1962, he addressed the Congress in Washington where he posed his astronaut friends as pioneers of "this new age." He predicted that after these flights in space, other men, younger and wiser would take the baton and run further. He commented on their spirit of teamwork and cooperation and closed his address with a statement on his faith in God:

> "We are proud to have been privileged to be part of this effort, to represent our country as we have. As our knowledge of the universe in which we live increases, may God grant us the wisdom and guidance to use it wisely."

I remember how proud I felt when we saw John Glenn escorting President John F. Kennedy's widow and children around immediately after the assassination. He was a father-hero to them as they attended the funeral and burial of the President. TV cameras showed him in close attendance with the family. What a blessing he must have been to them as the days went by after that terrible tragedy. That act was a great service to our country as all of us grieved at the loss of our President.

John Glenn has served as a U.S. Senator from Ohio for many years and has been in a position of respect from family and friends alike during his tenure in public office.

Now John wants to end his service as a Senator and offer his service to the elderly of the world if NASA and the National Institute of Aging could send him this time into space aboard the Shuttle to aid in medical research for the elderly. This announcement was made by Gannett News Service on July 20, 1997. It was reported that he had been in touch with NASA for a long time. This is truly a John Glenn idea; truly a national hero for today.

M. SCOTT CARPENTER, Commanded Mercury-7 flight in a three-orbit flight of the Earth on May 24, 1962. (NASA Photo)

"M. SCOTT CARPENTER was born in Boulder, Colorado, May 1, 1925. After World War II he entered the University of Colorado and received a bachelor of science in aeronautical engineering.

Carpenter participated in the V-5 flight training program sponsored by the United States Navy at the University of Colorado in 1943.

Carpenter was one of the seven astronauts selected by NASA in April 1959. He was backup pilot to John Glenn on the Mercury-Atlas 6 flight. Carpenter commanded the MA-7 three-Orbit flight on May 24, 1962, and was awarded the NASA Distinguished Service Medal for his space voyage. Later he resigned from the astronaut program to return to the Navy and pursue his interest in deep sea exploration." (28)

An assurance that Carpenter depended upon God was given in his portion of the book, WE SEVEN, Simon and Schuster, New York, N.Y., page 344. I quote: "I opened the hatch on the small end of the capsule, put the camera I had been using in a safe place near the opening and dropped the life raft into the water. I got into it before I realized that it was upside down. I climbed out into the water, turned the raft over and got back in. Then I tied the raft to the capsule so we wouldn't drift apart, and turned on the SARAH beacon which would help the recovery aircraft home on my position. And then I said a prayer, 'Thank you, Lord,' and relaxed for the wait."

This statement establishes Scott's dependence upon God to rule in his life.

L. GORDON COOPER, JR. Piloted Mercury-Atlas 9 on a 22-Orbit Mission In May, 1963. (NASA Photo)

"L. GORDON COOPER, JR. was born March 6,1927, in Shawnee, Oklahoma. He was graduated from the Air Force Institute of Technology, Wright-Patterson Air Force Base, Ohio, in 1956 with a bachelor of science degree in Aeronautical Engineering.

Cooper entered the United States Marine Corps in 1945 and later attended the Naval Academy Preparatory School. He was a member of the Presidential Honor Guard in Washington until his discharge in August 1946. Cooper entered the University of Hawaii, Honolulu, and received an Army commission after attending for three years. He transferred this commission to the Air Force and entered flight training in 1949. Upon completion of this training, Cooper flew with the 86th Fighter Bomber Group in Munich, Germany, for four years. After graduating from the Air Force Institute in 1956, he was assigned to the Air Force Experimental Flight Test School at Edwards Air Force Base, California.

Cooper, one of the seven astronauts selected in April 1959, on May 15, 1963, piloted MA-9 spacecraft on a 22-orbit mission. Cooper was command pilot for the Gemini V flight launched from Cape Kennedy, August 21, 1965. He was also assigned as command pilot of the back-up crew for Gemini XII.

He was awarded the NASA Distinguished Service Medal and the NASA Exceptional Service Medal."(29)

Gordon Cooper came from a long background in the Methodist Church. Throughout his service career, he and his family were active in the chapel services or in local Methodist churches. After his historic 22-orbit flight, he was awarded the World Methodist Council Medallion. At one of the World Methodist Council meetings, he told the over-400 present that he went through "stacks and stacks" of paper trying to write a suitable prayer before his space trip and finally gave up. He told those gathered that a prayer came to him spontaneously during the 17th orbit of the flight and he spoke it into the voice recorder as follows:

"Father, thank you, especially for letting me fly this flight. Thank You for the privilege of being able to be in this position; to be up in this wonderful place, seeing all these many startling, wonderful things that you have created. Help guide and direct all of us that we may shape our lives to be better Christians, trying to help one another rather than fighting and bickering. Help us to complete this mission successfully. Help us in our future space endeavors that we may show the world that a democracy can compete, and still are able to do things in a big way, and are able to do research, development, and can conduct many scientific and technical programs.

Be with our families. Give them guidance and encouragement, and let them know that everything will be okay. We ask in Thy name. Amen."

Several media sources reported how he came to name his capsule, "Faith 7." There were three reasons, he said: "One, because I believe in God and country; two, because of the loyalty to organizations, to the two organizations, actually, to which I belong, and three, because of the confidence in the entire space team."

EDWARD H. WHITE, II, spent 21 minutes outside of Gemini 4 Spacecraft (First man to perform extra-vehicular activity)June 1965 (NASA Photo)

"EDWARD H. WHITE, II was a West Point graduate and America's first spacewalker. He was among the second batch of astronauts named in September, 1962, and as Gemini 4 pilot in June, 1965, spent 21 minutes outside his spacecraft to become the first man in history to control himself in space with a maneuvering unit during extra-vehicular activity

Ed White was the only member of the first Apollo crew to have completed his work for a master's degree. His degree was in aeronautical engineering from the University of Michigan after his graduation from West Point.

He was a native of San Antonio and the son of a retired Air Force general.

Other highlights of his Gemini 4, four-day mission of 66 earth revolutions, included cabin depressurization, opening of cabin doors, and 12 scientific and medical experiments - all of them space "firsts."

His decorations and honors include: Golden Plate Award, NASA Exceptional Service Medal and Senior Astronaut Wings, Medalha Banderiantes va Cosmanautica, Firefly Club Award' one of the ten Outstanding Young Men of America in 1965; one of the five Outstanding Young Texans the same year, and the John F. Kennedy Trophy."(30)

Ed White gave us a glimpse into his faith in God with his favorite prayer as printed in the October 1965 issue of Guideposts:

> "Oh God, our Father, Thou searcher of men's hearts, help us to draw nearer to Thee in sincerity and truth...Make us to choose the harder right instead of the easier wrong, and never to be content with a half truth when the whole can be won.
>
> Endow us with courage that is born of loyalty to all that is noble and worthy, that scorns to compromise with vice and injustice, and knows no fear when truth and right are in jeopardy...
>
> Kindle our hearts in fellowship with those of a cheerful countenance, and soften our heart with

sympathy for those who sorrow and suffer....

Help us, in our work and in our play, to keep ourselves physically strong, mentally awake and morally straight, that we may better maintain honor of the corps untarnished....to realize the ideals of West Point in doing our duty to Thee and to our country, Amen." (31)

Like Gordon Cooper, Ed White was a member of the Methodist Church. At my church, First United Methodist Church of Cocoa, Florida, we memorialized Edward White in our "Dominion Window," a beautiful 11 X 22 foot faceted-glass window which displays Ed and the Gemini spacecraft in his famous spacewalk. Every time I look at it, my eyes never fail to take in that beautiful scene he gave us on that occasion. The window was designed around the 8th Psalm which several astronauts have quoted, and much of its design was inspired by the astronautical and scientific things that were taking place at the Cape.

I had served on the building committee for my church, and it was my privilege to suggest this particular segment for the window in memory of astronaut White who gave his life for the Apollo Program in the command module fire on January 27, 1967.

FRANK BORMAN, flew two Gemini Flights and was pilot of Apollo 8, First to visit the vicinity of the moon. December, 1968 (NASA Photo)

"FRANK BORMAN was born in Gary, Indiana, March 14, 1928. Raised in Tucson, Arizona, he graduated from the United States Military Academy in 1950 with a bachelor of science degree.

After receiving pilot training at Williams Air Force Base, Arizona, he was assigned to various fighter squadrons in the United States and the Philippines. In 1957 he became an instructor in thermodynamics and fluid mechanics at the military academy.

Borman received a master's degree in aeronautical engineering from the California Institute of Technology in 1957. He was graduated from the USAF Aerospace Research Pilots School in 1960 and remained there as an instructor until 1962.

Borman was one of the nine astronauts selected by NASA in September 1962. He was command pilot for the Gemini VII mission, which established a number of records, including the longest manned space flight performed up to that time (330 hours and 35 minutes). He participated in the first rendezvous to two manned maneuverable spacecraft (Gemini VI and Gemini VII) on December 15, 1965.

Borman has been awarded the NASA Exceptional Service Award and was co-recipient of the 1966 Harmon International Aviation Trophy for his flight achievements." (32)

At the 156th annual meeting of the American Bible Society in May 1972, Frank gave a remark about his belief in the Holy Bible:

> "I hear all over that the Bible does not relate, that it belongs to other places, other times. The fact is, it was never more applicable. It does ask to be met halfway, and it does demand analytical study. In return it continues to reveal truths set down for a less sophisticated generation of mankind, but the seeker will find that the principles in these truths will remain timeless. The Bible contains words that are both a

promise and a mandate. But more than anything they are an answer."

On December 21, 1968, astronauts Borman, Lovell, and Anders set sail for the moon from Kennedy Space Center on Apollo 8. Somewhere during their ten orbits of the moon they gave all the peoples of the Earth a Christmas present. At the time of the launch I did not know anything about the faith of these three astronauts. I had seen them around a few times and at meetings, but did not know much about their backgrounds. I did know that Frank Borman was a lay leader at St. Christopher Episcopal Church at League, Texas, near his home in the Houston area. You can imagine my excitement when on Christmas Eve we heard the three of them read the creation story from the Book of Genesis. (Later, NASA Public Affairs provided a tape to me of that portion of their transmission from the vicinity of the moon). Borman was the last to read, and at the end, made this personal statement: "And from the crew of Apollo 8, we close with good night, good luck, a Merry Christmas and God bless all of you - all of you on the good Earth!"

In briefing the Subcommittee on Manned Space Flight on 28 February, 1969, Dr. Kurt Debus, Director, Kennedy Space Center said of the flight:

> "The impact of Apollo 8, in my opinion, is something that defies quantitative measurement. Following the launch of Apollo 8, the Kennedy Space Center received over 5000 telegrams, phone calls, and letters from all over the world. This was by far the greatest volume of messages following a launch. A similar theme ran through all of these communications. For Americans, it was one of intense pride in their achievements. From friends all over the world - and letters came from 28 countries - it was one of pride in the human race and a feeling of gratitude to America. These letters came from men, women, the elderly, the young, the black, the yellow, the Christian, the Jew, the

Moslem, the Heads of State, the laborer, the engineer and the under-privileged. Apollo 8 had some specific significance for everyone." (33)

Borman's special prayer for the Apollo 8 mission was published in the media press as follows:

> "Give us, O God, the vision which can see Thy love in the world in spite of human failure.
>
> Give us the faith to trust the goodness in spite of our ignorancc and weakness.
>
> Give us the knowledge that we may continue to pray with understanding hearts, and show us what each one of us can do to set forward the coming of the day of universal peace. Amen."

As one looks at Dr. Debus' report and reads Borman's prayer, you cannot fail to see God's hand of providence and abiding love in the whole Apollo 8 mission. It was the first time mankind had made such a bold step in taking dominion in space. We find in Revelation 4:11, a statement which I trust, was fulfilled in this mission:

> "Thou art worthy, O Lord, to receive glory and honour and power: for thou hast created all things, and for thy pleasure they are and were created." (Revelations 4:11)

Can we hope that the human race, in this seemingly joint venture, gave God a bit of pleasure?

THOMAS P. STAFFORD, Command Pilot of Gemini VII;
The First Crew to Achieve Rendezvous in Space
December 15-16, 1965 (NASA Photo)

"THOMAS P. STAFFORD was born in Weatherford, Oklahoma, September 17, 1930. He received a bachelor of science degree from the U.S. Naval Academy in 1952.

He was commissioned in the Air Force and completed flight training and advanced interceptor training. He served five years as a fighter-interceptor pilot and subsequently was graduated from the Air Force Experimental Flight Test School in 1959. He is coauthor of Pilot's Handbook for Performance Flight Testing, and Aerodynamics' Handbook for Performance Flight Testing.

Stafford was one of the nine astronauts selected by NASA in September 1962. On December 15-16, 1965 with astronaut Walter M. Schirra, Jr., as command pilot, he was pilot of the Gemini VI flight, which established a space "first" by achieving rendezvous with the orbiting Gemini VII.

Stafford was command pilot of the Gemini IX mission on June 3, 1966, during which pilot Eugene A. Cernan spent two hours and seven minutes in extravehicular activity, and multiple rendezvous was achieved using three different techniques.

Stafford has twice been awarded the NASA Exceptional Service Medal and received the American Astronautical Society Flight Achievement Award in 1966. He was also awarded the 1966 Harmon International Aviation Trophy." (34)

Brig. General Stafford continued in the space program and was later assigned to the Apollo-Soyuz mission as the command pilot. His team-mates were Vance Brand and Donald (Deke) Slayton. Stafford, who attended the Seabrooke United Methodist Church in Houston, Texas, spoke at Oral Roberts University after his flight and stated his faith in God when he said, "If I did not believe in God, I would not get within 500 miles of Cape Kennedy." Also:

"In working with the Russian cosmonauts we have developed a very close friendship, despite the fact that the philosophies of the two countries are so opposite," he said.

On July 26, 1990. I was invited to attend the fifteenth-year reunion of the Apollo-Soyuz team. Tom Stafford was his usual fatherly self as he hosted General Aleksy Leonov and Valery Kubasov at the Kennedy Space Center. Deke Slayton and Vance Brand were unable to join their crewmates on this occasion, but all that attended enjoyed the reunion as each of the three members said a few words. My memory recorded the keen feelings they had for each other and for more cooperation between our two countries in future space ventures. It was a pleasure to shake each hand and wish them God's blessings during their visits to the United States, and particularly to the other NASA Centers.

Tom Stafford, in 1991, was chosen to head the Space Exploration Initiative, a challenge for America in the 21st century in space. The comprehensive report was submitted on schedule and was well received by the Bush administration and the Congress at the time. Tom's committee sent me a copy of the Initiative report which I cherish because he was asking us members of the NASA Alumni League to participate in the dialogue and review effort.

ALAN L. BEAN, Apollo 12, The Second Manned Expedition to The Moon, November 14-24, 1969 (NASA Photo)

"**ALAN L. BEAN**, born in Wheeler, Texas, March 15, 1932 was graduated from the University of Texas in 1955 with a bachelor of science degree in aeronautical engineering.

After graduation, Bean was commissioned in the United States Navy. He received Pilot Training and was assigned to Attack Squadron 44 at Jacksonville Naval Station for four years. He then attended the Navy Test Pilot School at Patuxent, Maryland, and served as project officer there in various aircraft for Navy preliminary evaluation, initial trials, and final board of inspections and survey trials. Bean's last Navy assignment was with the Attack Squadron 172 in October 1963, at Cecil Field, Florida.

Bean was selected by NASA as one of the 14 astronauts chosen in October 1963. He was command pilot for the backup crew for the Gemini X mission." (35)

Alan was assigned useful tasks in the space program as the months went by before he was assigned to Apollo 12 as Lunar Module Pilot. Apollo 12 sustained two direct lightning strikes as it rose into the air over the Atlantic Ocean, resulting in much excitement to the crew as power was lost and then restored automatically by the backup power source. They were able to go on with the mission to the moon and he and Charles Conrad, Jr. spent seven hours forty-five minutes setting up scientific instruments, collecting pieces from the nearby Surveyor spacecraft, gathering materials and photographing the Lunar Lander, the Surveyor and other objects of interest.

It was evident to us in the Launch Control Center that God did indeed protect our astronauts from lightning strikes. No other explanation for their safety was voiced. In collecting articles about Captain Bean, I came across the following item from the February 1970 United Methodist "TOGETHER" MAGAZINE:

"Navy Captain Alan Bean, the first United Methodist person to explore the moon carried a religious banner with him on his half-million-mile trip.

The banner, depicting the history of Christianity, including these religious symbols: the rose of Martin Luther, a crusader's cross representing Roman Cath-licism, an open Bible as a symbol of Calvinism, a fish as a symbol of the early church, an Anglican Chalice, a world and cross from the Anabapist tradition, a trifoil of the Trinity, and the United Methodist symbol of a black cross super-imposed over a red flame.

Captain Bean and his family are members of the Clear Lakc Unitcd Methodist Church of Clear Lake City, Texas, near Houston. The banner was presented to the church after the astronauts were released from quarantine.

The Reverend John E. Fellers, Clear Lake pastor, described the astronaut as "very interested in relating his religious convictions to his everyday life in terms of his actions and conduct toward his fellow man."

Prior to the launch, Captain Bean said that two things, major factors in a man's life, were important as he got ready to venture into space - one was his technical knowledge and ability and the other was his religious convictions." (36)

Who knows but what God, in His infinite mercy, spared the Apollo 12 crew from the lightning strikes just because of the religious banner that was to be used back on earth to honor Himself.

One year, Al Bean was our keynote speaker at our Range And Space Pioneer's banquet at Cocoa Beach where he impressed us with samples of his artistic creations. Such great artistic talent as his is unique and God-given as he shows the world through his paintings what it was like to walk on the moon.

EDWIN E. (BUZZ) ALDRIN, JR. Second man to step on the Moon, July 20, 1969. (NASA Photo)

"EDWIN E. (BUZZ) ALDRIN, JR. was born in Montclair, New Jersey, January 20, 1930. He graduated from the United States Military Academy in 1951 with a bachelor of science degree.

A member of the United States Air Force, Aldrin completed his pilot training at Bryan, Texas in 1952. He flew combat missions during the Korean conflict; he also served a tour as aerial gunnery instructor at Nellis Air Force Base, Nevada, and attended Squadron Officer's School at the Air University, Maxwell Air Force Base, Alabama. Aldrin flew as a flight commander with the 36th Tactical Fighter Wing at Bitburg, Germany.

Aldrin completed work for his doctor of science degree in astronautics at Massachusetts Institute of Technlogy in 1963 and was assigned to the Gemini Target Office of the Air Force Space Systems Division, Los Angeles, California. He was later assigned to the Manned Spacecraft Center, Houston, Texas, to coordinate Department of Defense participation in the Gemini experiments.

Buzz was one of the 14 astronauts selected by NASA in October 1963. He was pilot of the backup crew for the Gemini IX mission and pilot for the Gemini XII flight. He has received the NASA Exceptional Service Medal for his space flight achievements."(37)

Aldrin continued in the space program through various assignments and was eventually selected as Lunar Module Pilot for Apollo 11, the first lunar landing mission. On 16 July, 1969, Buzz, Neil Armstrong, and Michael Collins set sail for the moon where he and Neil spent 22 hours on the lunar surface, while Michael circled the moon in their command module.

The Aldrin family worshiped at the Webster Presbyterian Church just outside Houston, Texas where Pastor Dean once made this statement: "One of the principal symbols is that God reveals himself in the common elements of everyday life."

Buzz thought back to this statement while he was

working on the sophisticated tools of the space effort and related them as typical elements of life today. Then he wondered if he might take communion on the moon symbolizing the thought that God was revealing Himself there too, as man reached out into the universe. Buzz also made this statement when he wrote of this experience: "For there are many of us in the NASA program who do trust that what we are doing is part of God's eternal plan for man."

One of my friends gave me a 8 1/2 x 14 inch copy of Buzz's story, and the Orlando Sentinel of February 24, 1971 printed the same story, but I took this account from my handout:

> "Now Neil and I were sitting inside Eagle, while Mike circled in lunar orbit, unseen in the black sky above us. In a little while after our scheduled meal period, Neil would give the signal to step down the ladder onto the powdery surface of the moon. Now was the moment for communion.
>
> So I unstowed the elements in their flight packets. I put them and the scripture and the scripture reading on the little table in front of the abort guidance system computer.
>
> Then I called NASA headquarters in Houston: "Houston, this is Eagle. This is the LEM Pilot speaking. I would like to request a few moments of silence. I would like to invite each person listening in, wherever and whomever he may be, to contemplate for a moment the events of the past few hours and to give thanks in his own individual way."
>
> For me this meant taking communion. In the radio blackout, I opened the little plastic packages which contained bread and wine. I poured the wine into the chalice our church had given me. In the one-sixth gravity of the moon the wine curled slowly and gracefully up the side of the cup. It was interesting to think that the very first liquid ever poured on the moon, and the first food eaten there, were communion elements.

And so, just before I partook of the elements, I read the words which I had chosen to indicate our trust that as man probes into space we are in fact acting in Christ.

I sensed especially strong my unity with our Church back home, and with the Church everywhere. I read: "I am the vine, you are the branches. Whoever remains in me, and I in Him, will bear much fruit; for you can do nothing without me."

In his book, "RETURN TO EARTH," Buzz told this story and added that he had intended to send the communion message back to earth, but that NASA advised against it because of their legal battle with Madelyn Murray O'Hare, the celebrated opponent of religion.

As Buzz was performing his divine ritual in the LEM, Mike Collins was being reminded of his own faith in God there in the Command Module as he circled the moon and lent his support to the whole Apollo 11 mission as no one could do better. This is how he felt God's part in the mission as he phrased it in his book, "CARRYING THE FIRE," Straus and Giroux, New York, N.Y., page 410:

"As I scurry about, blocking off the windows with metal plates and dousing the lights, I have almost the same feeling I used to have years ago when, as an alter boy, I snuffed the candles one by one at the end of a long service. Come to think of it, with the center couch removed, *Columbia's* floor plan is not unlike that of the National Cathedral, where I used to serve. Certainly it is cruciform, with the tunnel up above where the bell tower would be, and the navigation instruments at the altar. The main instrument panels span the north and south transepts, while the nave is where the center couch used to be. If not a miniature cathedral, then at least it is a happy home, and I have no hesitation about leaving its care to God and Houston as I fade away into a comfortable snooze." (38)

What a mighty witness to Christian faith Mike gave

us as he put this into writing for our building up. Thank you, Mike!!!!

In my collection of background data for this book, I ran across a scrap of a newspaper clipping where Rev. Dean David Collins had read, from “The Living Church,” the moving "Prayer for Astronauts" which follows:

"O Eternal God, whose heavens declare thy glory and whose firmament showeth thy handiwork; we commit to thy almighty protection Thy servants for whose preservation in the boundless reaches of outer space our prayers are offered. Guard them, we beseech thee, from the dangers of the void, from sickness, from the violence of the cosmic elements, and from every peril to which they may be exposed. Conduct them with confidence to the accomplishment of their mission and to a safe re-entry and return, with a grateful sense of Thy mercy; through Jesus Christ our Lord, who with the Father and the Holy Spirit, liveth and reigneth in the uttermost parts of Thy universe, and to ages of ages. Amen."

It is no wonder that Almighty God blessed this trip to the moon and return. His answer to this prayer was felt all over the world as the TV links kept mankind abreast of this great undertaking.

JAMES B. IRWIN, Lunar Module Pilot on Apollo 15 Mission. Was First to use The Lunar Rover, July 26- August 2, 1071 (NASA Photo)

"JAMES B. IRWIN was born in Pittsburgh, Pennsylvania, on March 17, 1930. He received a bachelor of science degree from the U.S. Naval Academy in 1951 and a master of science degree in aeronautical engineering and instrumentation engineering from the University of Michigan in 1957.

He was commissioned in the United States Air Force in 1951 and received flight training at Hondo Air Base, Texas, and Reese Air Force Base, Texas. Irwin has served with the F-12 Test Force at Edwards Air Force Base, California, and the AIM 47 Project Office at Wright-Patterson Air Force Base, Ohio. He graduated from the Air Force Experimental Flight Test Pilot School in 1963. His assignment prior to his selection as an astronaut was as chief of the Advanced Requirements Branch at Headquarters Air Defense Command.

Irwin was one of the 19 astronauts selected by NASA in April 1966." (39)

Colonel Irwin was a member of the astronaut support crew for Apollo 10 and backup lunar module pilot for Apollo 12. On the Apollo 15 mission (July 26 to August 7, 1971), he served as Lunar Module Pilot. His crew was the first to use the lunar rover vehicle. While Colonel Al Warden orbited the moon in the Command Module, Irwin and Apollo Commander, Colonel Dave Scott, explored Hadley Rille and the Apennine Mountains of the moon, and they collected 180 pounds of lunar surface material including the Genesis Rock. On the flight of Apollo 15, Irwin logged 295 hours and 11 minutes in space, of which 19 hours and 46 minutes were spent on the moon's surface. Colonel Irwin, a member of the Nassau Bay Baptist Church in Houston, was the first man to quote scripture on the surface of the moon (Psalm 121:1). (Buzz Aldrin would have been the first if he had been allowed to read the traditional communion service from the Book of John).

Irwin is the author of "TO RULE THE NIGHT," his autobiography and personal statement of faith. His book, "DESTINATION MOON, " was his story of Apollo 15 and experiences on the moon. His "MORE THAN EARTHLINGS," and "MORE THAN AN ARK ON ARARAT," comprise the story of his search for Noah's Ark.

Special honors awarded Jim Irwin include the NASA Distinguished Service Medal, Air Force Distinguished Service Medal, The United Nations Peace Medal, the Robert J. Collier Trophy, The George Washington Medal of Freedoms Foundation, the Military Chaplains Association's National Citizenship Award, and numerous other military, community and service awards.

I attended a prayer breakfast on February 5, 1986 at Melbourne, Florida in order to hear Jim give his story of the Apollo 15 mission. He told us that something happened to him that he would never forget. He said, "It wasn't long (after landing on the moon) before I sensed that the Lord was there. I never saw Him, but I was aware of His presence." He told our crowd of 300 how an unseen voice helped him repair a malfunctioning science station they were erecting on the moon's surface. "I was standing there perplexed, and in my confusion I said, 'Lord, what should I do?' He heard these words, "Jim, down on your knees!"

"So I just sank to my knees there by this package on the surface of the moon, and from that position I could see what had broken," he said. "Then even with those very thick gloves on, I was able to insert my fingers and remove the part that had failed. The station was released and it just erected itself dramatically right in front of me like a small altar."

One of his final statements was that: "God loves us and wants to share our lives. We are dependent upon God. I have seen the Earth as God sees it. We are special in the universe. We have had God's help and direction in the space program."

Another time that God helped Jim was on the day they set out on the lunar rover to collect more moon rocks for the geologists back home. The March 1972 issue of GUIDEPOSTS gave this account:

"Through all of this, I never lost the feeling of the

closeness of God. Especially during the so-called "evening" of the moon, when we were bedded down and quiet, I felt that same anticipation I had known as early as the day we made a prayer pact with the church back home. "Please, Lord," I prayed, "help us have a rewarding exploration tomorrow."

So it was that the next morning Dave and I took the rover once again across the lunar surface, leaving our tracks behind us.

We came to a crater, old and badly mauled by millions of years' bombardment with meteorites. We slowed down the rover, watched the harness flaps settle back with the peculiar slow-motion effect of movement on the moon, then we stopped. We dismounted and here at Spur Crater we started our exploration across the dust.

And there it was.

Dave and I saw it at almost exactly the same time. For we could not possibly have missed it.

Incredibly, one peculiar-looking rock sat on the edge of the crater on a sort of pedestal, almost as if a hand were holding it up to us. Even before we reached out to pick up the rock we realized that it was quite different from anything else we had seen. It was crystalline, not bubbly. It was not like the rock Dave had chopped off the block earlier. It was in fact not like any rock we had ever seen before. Long inclusionlike streaks ran down the side. And they were white.

Being on the pedestal was remarkable enough, but it wasn't until we were about to reach out and pick up the rock that we realized the most astonishing thing of all.

For although the floor of that crater was deep in dust, and although the hand-like pedestal that held our rock was itself deep in dust - the sample of mineral we were staring at had no dust at all!

Slowly, Dave picked the rock up off its pedestal and turned it over. The underside was pure white!" (40)

One of the primary reasons for going to the high

Apennine Mountains and to the Hadley Rille area was to learn about the origins of the earth by studying the moon. They both had similar histories. Both were formed some four and a half billion years ago, but our earth has seen centuries of erosion by wind and rain while the moon has only been changed by falling meteorites. The name, "Genesis Rock" had already been coined for such a rock, so God in such a miraculous way allowed the Apollo 15 crew to find it because they had asked for His help. I firmly believe that most of our feeble accomplishments in the space program would not have been accomplished without His help. Praise God!

In 1979 Jim organized his "High Flight Foundation," an evangelistic organization where another astronaut, William R. Pogue, joined in this international outreach ministry. High Flight Foundation financed six expeditions to Turkey's Mount Ararat which Jim led, searching for the remains of Noah's Ark.

Jim Irwin died of a heart attack on August 9, 1991. He had a history of heart problems, but this did not stop him from declaring his faith wherever he could. He was a true American Hero and a great Christian adventurer. I am particularly grateful to him for encouraging me to publish this book. All of us who have been ministered to by Jim can say with the Psalmist, "Precious in the sight of the Lord is the death of His saints," Psalm 116:15).

CHARLES M. DUKE, JR., Apollo 16 Mission to The Moon, April 16, 1972. (NASA Photo)

"CHARLES M. DUKE, JR. was born in Charlotte, North Carolina, October 3, 1935. He received a bachelor of science degree from the United States Naval Academy in 1957 and a master of science degree in aeronautics and astronautics from the Massachusetts Institute of Technology in 1964.

He was commissioned in the United States Air Force in 1957, upon graduation from the Naval Academy. He completed flight training in 1958, and was later assigned to the 526th Fighter Interceptor Squadron at Ramstein, Germany, where he served three years as a fighter interceptor pilot. His last assignment prior to his selection as an astronaut was as an instructor in control systems at the Air Force Research Pilot School, from which he graduated in 1965. Duke was one of the 19 astronauts selected by NASA in April 1966." (41)

According to Charlie Duke's testimony in the GUIDEPOST magazine on July 1984, he did not take his faith in God into space with him as he winged on his way to the moon, but had a beautiful experience as he and Ken Mattingly prepared for their space walk to retrieve film canisters used during lunar orbit. This concerned Ken's wedding ring which had floated out of their sight within the cabin and later floated out the cabin hatch after them. Charlie grabbed for the ring knocking it into Ken's helmet. "In awe, we watched it bounce off Ken's helmet and float back to him," he said, "I grabbed it. We were all amazed. The odds for a disc tumbling through space to rebound perfectly off another object are practically zero."

He said, "Now I wonder if that odd experience was trying to tell me something. I looked at my own wedding ring. Was it something about my own marriage.?"

His testimony continued with details of his troubled marriage and how God began to work in his life through their friends as they witnessed to him about the Lord. A weekend Bible study retreat cleared it all up for Charlie, and today he is a most positive witness for the Lord Jesus.

Dick Baumbach's article in the August 11, 1979

Florida TODAY newspaper told how Charlie began this second exciting adventure which proved even more exciting than walking on the moon:

> ....He devoted much of his life to the Apollo space program. He holds the distinction of being the 10th man to walk on the moon.
>
> But now, former astronaut Charles Duke's days are taken up with talk not about space feats, but Jesus Christ.
>
> "I've found an inner peace that I never thought I would find. I found it once I came to the Lord," Duke said recently.
>
> That inner peace and the excitement which he says goes along with knowing "Jesus as my personal saviour" far surpasses the thrill of being an Apollo moon-walking astronaut.
>
> "Walking on the moon was a big accomplishment, but knowing Jesus just turns your life around," Duke, now a resident of San Antonio, Texas, explained.
>
> The soft-spoken Duke, now involved with real estate development, smiles broadly and seems very at ease as he talks about his new faith in Christ.
>
> He said it was a year and a half ago "when I was riding in my car going home from a Bible Study group that I realized that Jesus had all the answers.
>
> "I was with my wife Dotty and I turned to her and said everything just fell together for me as far as the Bible and Jesus was concerned," he said. Since that day, Duke who says he was a "Sunday Christian" before the experience, has been talking to everyone about his realization.
>
> "The Lord has a place for me and now I see it as telling people about Him," the former spaceman said.
>
> Duke accomplishes that goal when he gets invited to give talks about his Apollo Experiences.
>
> "When people find out you're a former astronaut they want to have you as a guest speaker. I usually

accept the invitations and then I get my opportunity to talk about Jesus.

I pray...(that) if the Lord wants me to talk about Him, he will have someone ask me a question about any religious experiences I had during the Apollo program."

"I almost always get asked a question about that topic. That's my opening to start talking about Jesus," he explained.

Duke says none of his audiences get upset with his preaching.

"In fact, people will come up to me after I finish speaking and they want to talk more about Jesus."

I've had people say, 'We're really glad you spoke about Jesus. That's what we wanted to hear.' "It makes me feel good that I can let others know that Jesus is real," he said. (42)

My decision to include Charlie in my book was prompted by Jim Irwin in his letter to me of November 7, 1990 where he witnessed that Charlie Duke should be included. He said, "You might want to include Charlie Duke who is just as active as I am in sharing his faith in Jesus Christ."

All of us need a friend like Jim Irwin.

Just as a note of encouragement to me, Jim signed his letter to me of November 7, 1990 as: "Your grateful brother from the moon, Jim Irwin, Apollo 15."

Thank you Jim, for telling me about Charlie's faith and for your encouraging compliments.

JACK R. LOUSMA, Participated In The Skylab Program and Flew on The Third Orbital Flight of Shuttle Columbia in 1982. (NASA Photo)

" JACK R. LOUSMA was born in Grand Rapids, Michigan, on February 29, 1936. He received a bachelor of science degree in aeronautical engineering from the University of Michigan in 1959 and the degree of "Aeronautical Engineer" from the U.S. Naval Postgraduate School in 1965.

He entered the United States Marine Corps in 1959 and completed flight training in 1960 at the U.S. Naval Air Training Command. He was assigned as an attack pilot, and subsequently served in Iwakuni, Japan.

Lousma's last assignment before his selection as an astronaut was as a reconnaissance pilot at the Marine Air Station, Cherry Point, North Carolina.

Lousma was one of the 19 astronauts selected by NASA in April 1966." (43)

Colonel Lousma participated in the Skylab Project on the second manned launch. His crew spent 59 days, 11 hours and 9 minutes in space on that mission. This was in 1973. In 1982 he spent 8 days in space on the third orbital flight on the space shuttle Columbia. He was in the back-up crew for the US-Russian Apollo-Soyuz Test Project and spent two months in the Soviet Union learning their language and then worked with them for two years. Later he was sent to the Peoples Republic of China and other nations.

The Full Gospel Business Men's publication, VOICE, for July 1985 allowed Colonel Lousma to make the following statement about his faith:

> "During all this time I was aware of God's leading and of his constant presence. While I lived in the Soviet Union I read the Psalms and Proverbs for my daily devotions. I looked up all the verses that related to the universe and the earth. There are many which extol God's universe and demonstrate that it is His. But one of the greatest is found in Psalm 19:1-6:"
>
> "The heavens are telling the glory of God; they are marvelous displays of His craftsmanship.

> Day and night they keep on telling about God. Without a sound or word, silent in the skies, their message reaches out to all the world. The sun lives in the heavens where God placed it and moves out across the skies as radiant as a bridegroom going to his wedding, or as joyous as an athlete looking forward to a race.The sun crosses the heavens from end to end, and nothing can hide from its heat." (The Living Bible)

We will never understand everything about our universe. There are enough galaxies for everyone in the world to own two or three of them personally. Yet this whole universe operates in perfect harmony and order, like a finely tuned watch. Only Almighty God could put it together and keep it running so perfectly.

As we learn more about the universe, we will come to know its Creator more fully and to appreciate Him in a greater way.

On March 22, 1982, I was at my console in the Launch Control Center when I heard Jack Lousma's voice on the intercom. He gave thanks for the best wishes and prayers of all the launch team at T-10 minutes. He also mentioned that the President had just called to wish the crew of STS-3 well. Lift-off came at 11:00 a.m.

The Florida TODAY newspaper on March 21, 1982 included a small item concerning Colonel Lousma's witness:

> "During 59 days aboard the Skylab space station in 1973, astronaut Jack Lousma often read passages from the Bible. Does he feel any closer to God when he's out there?
>
> "Anytime you're in an airplane or any kind of vehicle like that looking up, you have to feel somewhat closer to God, because you are detached from the earth to some degree," he replied in an interview.
>
> "But the answer, I think, is probably no," Lousma added. "Because my experience is that you can feel just as close to God anywhere on earth as you can feel in

> space. Sometimes people tend to use their lack of ability to get to great heights as a reason for not getting closer to God, and that's not a good excuse." (44)

I thank God for not only these statements from Jack, but for that moment in the countdown when he thanked us for our best wishes and for our prayers there at our consoles in the Launch Control Center.

WILLIAM R. POGUE, Participated in Skylab 4 Mission in 1973, The Longest Term Stay in Orbit (NASA Photo)

"WILLIAM R. POGUE was born in Okemah, Oklahoma, January 23, 1930. He received a bachelor of science degree in education from the Oklahoma Baptist University in 1951, and a master of science degree in mathematics from the Oklahoma State University in 1960.

He entered the Air Force in 1951 and received his commission in 1952. He served with the Fifth Air Force in 1953 and 1954 and flew 43 combat missions over Korea.

From 1955 to 1957 he was a member of the USAF Thunderbirds. Pogue has gained flying proficiency in over 50 types and models of American and British aircraft and is qualified as a civilian flight instructor. From 1960 to 1963, he served as mathematics instructor at the USAF Academy in Colorado Springs, Colorado. He is a graduate of the Empire Test Pilots School in Farnborough, England, and in 1965 he completed a two-year tour as a test pilot with the British Ministry of Aviation under the USAF/RAF Exchange Program. Prior to his selection as an astronaut, he was an instructor at the Air Force Aerospace Research Pilot School, Edwards Air Force Base, California.

Pogue was one of 19 astronauts selected by NASA in April 1966." (45)

On November 16, 1973, we sent astronauts Pogue, Carr, and Gibson on a course to rendezvous with Skylab, the United States first space station. Their's was the third crew to spend time in the station. Astronaut Pogue contributed to the success of the mission and was instrumental in preparing it for a long-term stay in orbit without a crew.

The Skylab 4 mission resulted in the longest-term stay in orbit and much useful information on man's tolerance of weightlessness was recorded for use in future long-duration flights. We owe this team a great deal of thanks for all the hardship and long hours they put in towards all their goals.

Exploits of the Skylab missions are recorded in the 1983 NASA publication, SP-4208, LIVING AND WORKING IN SPACE.

As Pogue and his crew prepared Skylab for its long-term stay in orbit, they envisioned a permanent space station with Skylab being the core element of many expansions if only congress could be persuaded.

The book NASA SP-4208, LIVING AND WORKING IN SPACE chronicled all the events that led to a decision to let Skylab decay in orbit and reenter the Earth's atmosphere and burn up.

Reboosting of the station by a planned Space Shuttle mission did not materialize, so after $3.75 million dollars were spent in the effort, Johnson Space Center Director, Chris Kraft, publicly expressed his opinion that the effort was futile. General opinion was that congress had not funded the Space Shuttle well enough to launch the reboost flight in time to rescue Skylab. Reentry occured on July 11, 1978 over Australia and ended the dream of many of us who wanted a permanent space station.

Astronauts Carr, Gibson and Pogue must have felt badly on February 8, 1974 when they moved into the command module and prepared for separation and return to earth, leaving their faithful but damaged space station in orbit to shift for itself until a Space Shuttle team could revisit.

Like some of his astronaut peers, William Pogue experienced a deeper commitment to God as the long hours in Skylab continued. He said that his voyage into space restored a "proper balance" to his outlook on life. He later resigned from NASA and joined High Flight Foundation, Jim Irwin's evangelistic organization, for a while, as a way to express his commitment to the Lord.

To me, the "proper balance" Astronaut Pogue mentioned, means that he had accepted Jesus as his personal Savior at some point along life's way. That act alone assured Pogue that he was not alone on his venture into space - that he had a constant companion in the person of the Holy Spirit to give him confidence, to assure him of life everlasting, and to give him faith and a quiet spirit wherever he was.

ROBERT SPRINGER, Flew As A Mission Specialist In Shuttle Discovery On March 13, 1989. (NASA Photo)

"ROBERT C. SPRINGER, born in St. Louis, Missouri, was a 1960 graduate of Ashland, Ohio High School. He is a graduate of the U.S. Naval Academy with a bachelor of science degree in naval science. In 1971 he received a master's degree in operations research and systems analysis from the U.S. Naval Postgraduate School.

Lt. Col. Springer was test pilot for more than 20 different types of fixed and rotary-winged aircraft and has logged more than 2800 hours of flying time in jet aircraft.

He flew 550 missions in Vietnam and was the aid-de-camp of the Commanding General, Fleet Marine Corps, and was responsible for joint operational planning for Marine forces in NATO and the Mid-East.

In May 1980, NASA selected him as a mission specialist. After his one-year training period he served as a member of the support crew for the third shuttle mission and was responsible for the Astronaut Office's coordination of reviews recertifying the Space Shuttle program after the Challenger accident.

Robert has received the Navy Distinguished Flying Cross, Bronze Star, and Combat Ribbon. He and his wife Molly have three children: Chad, Kira, and Derek. They attend the Presbyterian Church, and Robert is a member of the Full Gospel Men's Fellowship chapter at Johnson Space Center." (46)

Like a lot of the faithful, Robert had a gradual growth into a life with Christ. Quoting from the Full Gospel Business Men's "VOICE" of July/August 1983:

> "I knew now that God's hand had always been directing me even when I wasn't aware of it. I'd always considered myself a Christian, ever since child-hood when my parents sent me to Sunday school. But to seek God's guidance on a constant, day-by-day, minute-by-minute basis....well, I was just beginning to get the feel of that kind of Christianity......
>
> I got my wings in 1966 and went to Viet Nam the next year. I did a lot of flying with a fellow Marine, and

on our missions together we'd talk back and forth about God. Other people started overhearing our conversation and we got tagged with the nickname "Apostle Flight."

No matter what success or happiness comes my way, I know all the glory belongs to Jesus. Sometimes people ask me, "Bob, hasn't God always been a factor in your life?" Yes, He has. But I'm delighted to add that today He's something much more: He's a personal, loving, caring Friend." (47)

In 1987 Robert and Molly helped organize a church in Sea Brook, Texas with two other families. By early 1989, the Bay Area Presbyterian Church had more than 200 members and a great program for their youth in which they took great interest because of their concern for the children in the area.

Monday, March 13, 1989 marked a great day for Lt. Colonel Springer, as the years of training and preparation for space flight was rewarded when the Shuttle Discovery came to life at 9:57 A.M. Its mission was to launch a Tracking and Data Relay Satellite which provided a crucial link in the NASA communications system for future shuttles. As I watched the launch from my home, it was good to know that a fellow-Christian was present in the shuttle as I joined in prayer with him and the other crew members who took God's hand and went soaring into space that beautiful morning.

THE HONORABLE BILL NELSON, flew as a Mission Specialist on Shuttle Columbia, STS-61C, December 19, 1985 (Photo provided by Congressman Nelson's Staff).

THE HONORABLE BILL NELSON, flew as a mission specialist in Shuttle Columbia, STS-61C, on December 19, 1985. Bill was Chairman of the U.S. Congressional Space Science and Applications Subcommittee and felt that a trip into space would permit himself more insight into the needs and goals of the space program.

I have known for many years that Bill is a devoted Christian and is not ashamed to show his faith. I was not surprised to see confirmation of that when I read his book, MISSION, Harcourt Brace Jovanovich, New York, N.Y. The book was about his mission into space aboard Shuttle Columbia, co-authored by Rev. Jamie Buckingham of Melbourne, Florida. It was a "Report to the American people and their representatives in Congress."

Bill made notes on page 4 of his book that he had placed his pocket bible in his right pocket as part of his regular equipment.

On the morning of December 19th, he had been suited up and transported to the launch pad along with the astronauts. At the pad he entered the elevator and rose to the 195 foot level where they prepared to enter the Shuttle. On page 12 of MISSION we find this account of his next action:

> "Waiting in that cold, crisp, early dawn on the small steel grate walkway that extended from the launch tower to the door of the Shuttle, I stepped off to one side, away from the other men. To my right was the Atlantic Ocean. To my left was the state of Florida. I looked out over the expanse of the Cape, toward the land where my family had put down its roots. Stepping aside, I moved behind a large beam and knelt on the steel grate floor and mouthed a simple prayer, thanking God for this opportunity and asking Him to protect our mission."

Farther into his account of the launch preparations, he also stated that he found himself reciting the 23rd Psalm, which he had memorized as a boy. He also reported that at least one of the astronauts aboard was a man of

God, Charlie Bolden, who taught Sunday School in his Episcopal Church in Houston.

I am sure these two space-men were bolstering each other's faith and assurance that the flight would have God's blessing and go well.

After a successful flight and safe return to earth, Bill had this advice for the nation and the world on page 228 of his book:

> "Space exploration, like all pioneer work, demands more than physical courage and finances. It calls for a willingness to stretch the limits of one's horizons beyond the little worlds of comfort and fear in which we live - to include the broader spectrum of life. This means we explore not only space, but the possibilities of working with all those who share a common goal, or the grander purpose of "peace on earth to men of good will." (48)

In my research for this book, I have noticed a bright thread of continunity in God's act of placing key spokesmen into places of importance in His ongoing plans for we humans on Planet Earth. Surely Bill Nelson was one of those. Bill was ever faithful to our nation's space program as evidenced by his leadership in the U.S. Congress over his years of service. I remember how honored I was to meet him at many of the Space Congress programs at Cocoa Beach, Florida. Bill's keynote addresses were always upbeat, and you could just tell that he had the enthusiasm that it took to advance our space program into the heavens that we see taking shape today.

CONCLUSION

In conclusion, it seems to me that all of our astronauts have contributed immeasurably to just such a goal of "peace on earth to men of good will." I especially wanted to honor all the previously named astronauts for their relationship to God and for their understanding of His universe and their place in God's plan for mankind to take dominion of all that God created. I listed the astronauts in order of their initial flights into space, and have tried to share with the reader the faith they took into space as they braved the hostile elements of this final frontier which man has set out upon.

Today, as I write this conclusion, I must include some statements from the Associated Press on September 30, 1997 from the Florida TODAY newspaper, concerning the faith of astronaut David Wolf, the first Jewish American to go into space. His family reported that he took with him a mezuza, a rolled piece of parchment in a case that traditionally is attached to the doorpost of a new home. Well, the Space Station MIR will be a new home to Wolf for four months as he works in space with his two new Russian crewmates. This act of faith in God will be imputed to him for righteousness as he and his crewmates make the many repairs on that crippled outpost.

The prayer that is tucked inside the mezuza is: “Hear, O Israel, the lord our God, the lord is one” - an especially fitting line considering how borders and divisions are invisible from space. Astronaut Wolf has taken his faith into space and trusts God to be there as he expressed his confidence in the safety of the 11 year-old space station and the importance of his research while there.

There seems to be more honor and trust in God reported in the recent past, such as the the quoted prayer on page 82 reported by Chaplain Nguyen at Patrick Air Force Base where he had placed a prayer on each launch control console during launch countdown operations for the unmanned launches.

According to the Associated Press, the Russian Orthodox Church has started a praying tradition with Russian Astronaut launchings. A black-robed priest did the first-ever prayer service at the cosmonauts' hotel before their flight to the MIR space station which lifted off on Tuesday, August 5, 1997. This was to be a very dangerous mission for cosmonauts, Anatoly Solovyov and Pavel Vinogradov who had been in training for the extensive repairs planned at the space station after being damaged by a cargo supply ship on June 25, 1997.

One can see the hand of God at work in Russia as this one act of open praise and dependence on the one true God 's blessing on a State venture such as the Russian Space Program was inaugurated and announced to the whole world. Praise God and praise His many followers in Russia!

I am sure there are other astronauts who have a great faith in God and faith in His orderly universe who have not made a public statement. Their faith counts and their prayers are being answered as man continues to mount the sky and to seek God's face.

Wernher von Braun expressed it a little differently in his article to the TOGETHER magazine where he stated:

> "When space flight frees man from the final chains of gravity, it will open to him a new and wonderful experience of God's truth. It will open to him the gateway to heaven."

It is my desire that the present and future astronauts will be inspired and comforted in the knowledge of God's blessing on the space program as witnessed by those who went before.

I hope that this collection of testimonies from the space program will give a measure of "faith in space" to the world's astronauts and cosmonauts, and give them a greater faith in the orderliness of the universe they face at T-0, and as they work peacefully with other nations to open up this final frontier we call Space.

On the occasion of the first launch of the Space Shuttle's Spacelab, the Kennedy Space Center's "SPACEPORT NEWS" of December 9, 1983 printed the following message from the White House:

> "Each time we send a mission into space, we enter an area stretching millions of miles beyond the surface of the Earth. This effort is the focus of the hopes, dreams and daring of innumerable men and women who have made it the last great frontier of human endeavor.
>
> As we seek to conquer this frontier, we stand on the shoulders of giants of science out of the past whose scientific knowledge, like space itself, knows no international boundries. It is most fitting, then, that this ninth Space Shuttle mission, a true international cooperative endeavor, be devoted solely to the pursuit of increased scientific information in orbit.
>
> Spacelab will orbit the Earth in the Shuttle's cargo bay, is the first full-scale scientific laboratory to go into space. It was funded, developed, and built by the European Space Agency. I congratulate our European friends on this magnificent achievement. A dedicated team of European scientists and engineers worked ten years to complete it. The United States is proud to be a partner in what is the largest international space project in history.
>
> Spacelab will open new opportunities for scientific research, experimentation, and expanded industrial applications to better our lives on earth and allow us to do work in space to create jobs on our planet. In future years, we will look back on this mission marking the first flight of a non-American aboard the Shuttle as a milestone in our international cooperative efforts to use space for peaceful purposes benefiting all mankind.
>
> We send our best wishes to the crew of the Columbia as they set out on their journey. May Captain John Young and Pilot Brewster Shaw, Mission Specialists, Owen Garriott and Bob Parker, and Payload Specialists Ulf Merbold of ESA and Byron Lichtenberg all have a

successful mission. Bon voyage and God bless you."

Signed, Ronald Reagan

In this White House message to the Shuttle Columbia crew, one can see the dreams of many visionaries of the past for just such an international space program that President Reagan mentioned, especially President Kennedy, Dr. Wernher von Braun, Hermann Oberth, Konstantin Tsiolkovsky, Dr. Robert Goddard, former Deputy Administrator, Dr. Hugh L Dryden. and former NASA Administrator, Dr. Robert Frosch.

On page 36 Dr. Goddard was concerned that: "one day, far into the future the sun and its planets might grow cold, or lack life-sustaining resources for man's survival. At that time man might become universal explorers and seek more hospitable grounds where they might live and propagate the human race."

As I researched, wrote, edited, revised and finally completed this forty-plus year venture, I could not help seeing spaceflight's Godly heritage handed down through the centuries to we who carry the fire for our international space program. All the effort of so many men and women towards the goal of establishing the international space station, with attendant plans to venture to other planets, cannot help but succeed in taking dominion in space. May God continue to bless all of our international ventures.

FOOTNOTES

No.1: H. E. Mehrens, THE DAWNING OF THE SPACE AGE, Civil Air Patrol Publication.

No.2: February 1978, NASA ACTIVITIES NEWSLETTER.

No.3: March/April 1990, NASA ACTIVITIES NEWSLETTER.

No.4: Thomas H. Huxley, SCIENCE AND HEBREW TRADITION, D. Appleton and company, New York, N.Y, 1898.

No.5: Harry Rimmer, THE HARMONY OF SCIENCE AND SCRIPTURE, Reprinted by permission of W.B. Eerdmans Publishing Company, Grand Rapids, Michigan.

No.6: Dr. Owen J. Gingerich, CHRISTIAN HERALD Interview, December, 1978 issue. Permission granted by Christian Herald Magazine12-13-90.

No.7: Dr. Warren Weaver, CAN A SCIENTIST BELIEVE IN GOD from the book, SCIENCE AND IMAGINATION, Readers Digest, May 1968. (Reprinted with permission from Readers Digest, and fair use granted by Harper Collins Publishers. Copyright (c), Basic Books, Inc.)

No.8: A. Cressy Morrison, SEVEN REASONS WHY A SCIENTIST BELIEVES IN GOD, Readers Digest, October, 1960. (Reprinted with permission from Readers Digest, and from Fleming H. Revell company, taken from the book, *MAN DOES NOT LIVE ALONE*, Copyright A. Cressy Morrison (C). Fair use granted by Harper Collins Publishers.

No.9: DR. Eugene M. Emme, NASA Historian, ROBERT H. GODDARD, WORLD ROCKET PIONEER, NASA Historical Report No.1.

No.10: A.N. Petroff, KONSTANTIN EDWARDOVICH TSIOLKOVSKY, A Short Autobiography, Astronautics Magazine, May 1959 (Out of Circulation).

No.11: Helen B. Walters, HERMAN OBERTH, FATHER OF SPACE TRAVEL (Reprinted with permission of Simon & Schuster for Young Readers, an imprint of Simon & Schuster Children's Publishing Division from HERMANN OBERTH: FATHER OF SPACE TRAVEL by Helen B. Walters.

Copyright (c)1962 Helen B. Walters).

No.12: Florida TODAY Newspaper, January 22, 1990.

No.13: WERNHER von BRAUN, MISSILES AND CIVILIZATION, Together Magazine, October1959. (Out of circulation).

No.14: Steve Carman with Bob Owen, QUEST, Tyndale House, Publishers, Inc. Wheaton, Illinois. (Reprinted with permission of Mr. Steve Carman).

No.15: Major General J. Bruce Madaris, Retired, MY FLIGHT FROM DOUBT, Guideposts, September, 1968. (Reprinted with permission from Guideposts Magazine Copyright (c) 1968 by Guideposts, Carmel NY

No.16: Dr. Hugh L. Dryden, MAN'S ETERNAL QUEST, Together Magazine, January, 1962. (Out of circulation).

No.17: G.T. Willey, Orlando Evening Star, January 27, 1967.

No.18: Walter F. Burke, WAY OUT THERE, Decision Magazine, September, 1967. (Reprinted with permission of Mrs. Walter F. Burke).

No.19: Dr. Robert A. Frosch, NASA ACTIVITIES NEWSLETTER, December 1977 Issue.

No.20: Chris Kraft, Florida TODAY Newspaper, May 14, 1972.

No.21: Loyd S. Swenson Jr. , James M. Grimwood, Charles C. Alexander, THIS NEW OCEAN, NASA SP-4201.

No.22: July/August 1995, AD ASTRA MAGAZINE, National Space Soc.

No.23: NASA Public Affairs, NASA ASTRONAUTS, U.S. Government Printing Office, 1968, 0-290-288.

No.24: Florida TODAY Newspaper, April 20, 1986.

No.25: WE SEVEN, Simon and Schuster, New York, N.Y. (Reprinted with permission of Simon & Schuster from WE SEVEN by M. Scott Carpenter, L. Gordon Cooper, John H. Glenn, Jr., Wlater M. Schirra, Jr., Vrigil I. Grissom, Alan B, Shepard, Jr., Donald K. Slaton.)

No.26: Virgil Grissom, FLORIDA TODAY Newspaper, January 28, 1967.

No.27: NASA Public Affairs, NASA ASTRONAUTS, U.S. Government Printing Office,1968, 0-190-288.

No.28: Ibid

No.29: Ibid

No.30: Edward H. White II, Today Newspaper, January 28,1967, in part.

No.31: Guideposts Magazine, Edward H. White II, WHAT I TOOK INTO SPACE...by Astronaut Edward White, II originally appeared in Guideposts Magazine, October 1965. (Astronaut deceased.)

No.32: NASA Public Affairs, NASA ASTRONAUTS, U.S. Government Printing Office, 1968, 0-190-288.

No.33: Charles D. Benson and William B. Faherty, MOONPORT, A History of Apollo Launch Facilities and Operations, U.S. Government Printing Office, NASA SP-4202.

No.34: NASA Public Affairs, NASA ASTRONAUTS, U.S. Government Printing Office, 1068, 0-290-288.

No.35: Ibid

No.36: Alan Bean, RELIGIOUS BANNER MADE MOON TRIP, Together Magazine, February 1970. (Not in circulation).

No.37: NASA Public Affairs, NASA ASTRONAUTS, U.S. Government Printing Office, 1968, 0-290-288

No.38: Excerpt from CARRYING THE FIRE: AN ASTRONAUT'S JOURNEY by Michael Collins. Copyright (c) 1974 by Michael Collins. (Reprinted by permission of Farrar, Straus and Giroux, Inc.)

No.39: NASA Public Affairs, NASA ASTRONAUTS, U.S. Government Printing Office, NASA SP 4202.

No.40: James B. Irwin, MISSION TO THE HEART OF CREATION, Guideposts, March 1972. (Reprinted with permission from Guideposts Magazine. Copyright (c) 1972 by Guideposts, Carmel, NY

No.41: NASA Public Affairs, NASA ASTRONAUTS, U.S. Government Printing Office, 1968, 0-290-288.

No.42: Florida TODAY Newspaper, August 11, 1979

No.43: NASA Public Affairs, NASA ASTRONAUTICS,

U.S. Government Printing Office, 1968, 0-290-288.

No.44: Florida TODAY Newspaper, March 21, 1982

No.45: NASA Public Affairs, NASA ASTRONAUTICS, U.S. Government Printing Office, 1968, 0-290-288.

No.46: Robert C. Springer, Full Gospel Business Men's VOICE Magazine, Vol 31, No. I, July/August, 1983, Page 3-6, and the Florida Today Newspaper, March 11,

No.47 Ibid.

No.48: U.S. Congressman Bill Nelson with Jamie Buckingham, MISSION, Harcourt Brace Jovanovich, New York, N.Y. 1988. (Reprinted with permission of Bill Nelson).

Order Form

To Order additional copies of *Faith In Space*

Name:____

Address:______________________________________

City:____________________State:_________________

Zip:_____________Phone:_______________________

Please remit $ 14.95 per book: __________________

Sales Tax: Please add 6% for books
shipped to Florida addresses: ____________

Shipping: $3.00 for first book and
$1.00 for each additional book
shipped to the same address: ____________

Total: ____________

Mail to: Consecrated Press
P.O.Box 542346
Merritt Island, FL 32953

Order Form

To Order additional copies of *Faith In Space*

Name:______________________________________

Address:____________________________________

City:__________________State:_______________

Zip:____________Phone:_____________________

Please remit $ 14.95 per book: ________________

Sales Tax: Please add 6% for books
shipped to Florida addresses: __________

Shipping: $3.00 for first book and
$1.00 for each additional book
shipped to the same address: __________

Total: __________

Mail to: Consecrated Press
P.O.Box 542346
Merritt Island, FL 32953